MODELLING APPROACHES TO UNDERSTAND SALINITY VARIATIONS IN A HIGHLY DYNAMIC TIDAL RIVER

the case of the Shatt al-Arab River

MODELLING APPROACHES TO UNDERSTAND SALINITY VARIATIONS IN A HIGHLY DYNAMIC TIDAL RIVER

the case of the Shatt al-Arab River

DISSERTATION

Submitted in fulfilment of the requirements of
the Board for Doctorates of Delft University of Technology
and of the Academic Board of the UNESCO-IHE
Institute for Water Education
for the Degree of DOCTOR
to be defended in public on
Monday, 21 November 2016, 15:00 hours
in Delft, the Netherlands

by

Ali Dinar ABDULLAH
Master of Science in Water Resources Management, UNESCO-IHE
born in Missan, Iraq

This dissertation has been approved by the supervisors

Prof.dr.ir. P. van der Zaag   UNESCO-IHE / TU Delft
Dr.ir. I. Popescu   UNESCO-IHE / TU Delft
Dr.ir. U.F.A. Karim   University of Twente

Composition of the doctoral committee:

Chairman   Rector Magnificus TU Delft
Vice-Chairman   Rector UNESCO-IHE
Prof.dr. ir. P. van der Zaag   UNESCO-IHE/TU Delft, promotor
Dr.ir. I. Popescu   UNESCO-IHE/TU Delft, copromotor
Dr.ir. U.F.A. Karim   University of Twente, copromotor

Independent members:

Prof.dr. N. Al-Ansari   Lulea University, Sweden
Prof.dr. F. Martins   University of Algarve, Portugal
Prof.dr.ir. H.H.G. Savenije   TU Delft
Prof.ir. E. van Beek   University of Twente
Prof.dr. W.G.M. Bastiaanssen   TU Delft, reserve member

CRC Press/Balkema is an imprint of the Taylor & Francis Group, an informa business

Published by:
CRC Press/Balkema
PO Box 11320, 2301 EH Leiden, the Netherlands
Pub.NL@taylorandfrancis.com
www.crcpress.com – www.taylorandfrancis.com
ISBN 978-1-138-62625-6 (Taylor & Francis Group)

SUMMARY

Densely populated delta regions in hot climates are vulnerable to acute water availability and water quality problems, problems that are often interrelated. One of the biggest threats to water quality in such areas is salinity. Declining freshwater inflows of deteriorating quality are among the major issues faced by many such deltaic rivers and their dependent human- and ecosystems, notably in their most downstream areas, which are close to the major salinity sources from seawater, intensive farming, industrial and population hubs. The more surface water is diverted to fulfil demands for freshwater, especially during long drought periods, and the lower the quality of water that is returned to the main stream, the more inland the tidal excursion and seawater intrusion will penetrate. Water insecurity and ecosystem decline result with major socio-economic implications. Some of the immediate causes and effects of salinity may be evident by the naked eye but the full picture is less obvious in terms of longer-term consequences and strategies for remediation.

The process of salinization in rivers is dynamic and complex. It is multi-variate and can be highly variable, even random given the many factors and uncertainties associated with it. Determining these factors is data-intensive and case-specific and should therefore be tailor-made to past and current prevailing field conditions of salinity and a delta's hydrology and water use. Salinity studies in complex deltaic systems require therefore reliable historic data based on systematic field monitoring, preferably covering all possible sources of water salinity. This is essential for developing verifiable analytical and numerical models that are based on field-calibrated parameters that are system-specific. With appropriate monitoring and modelling of sufficient refinement rational diagnostic and policy decisions can be made for mitigating both water scarcity and salinity, which ultimately should lead to sustainable water development.

The present research is the first systematic monitoring and modelling study on water availability, water quality and seawater intrusion of the Shat al-Arab River (SAR). The SAR runs partially along the disputed international Iraq-Iran border, which has been a source of conflict in the past. It is an oil-rich area that has been continuously populated and cultivated since the beginning of civilization, characterised by the world's largest marshes and date-palms' forests. The Tigris and Euphrates rivers on the Iraqi side, and the Karkheh and Karun rivers originating in Iran, constitute the main water sources for the river and have allowed the region's rich ecological, socio-economic and cultural heritage. The river mouth on the Gulf is also an important shipping route on which the densely-populated Iraqi city of Basra with its huge ports and oil export terminals are situated. The present research is timely and relevant for the issues of origin, regulation, and management of salinity in the SAR have become highly politicized and are hotly-debated at provincial and national levels.

Current scientific knowledge on the SAR salinity problem is deficient, partially due to the complex, dynamic, and spatiotemporal interaction between salinity sources with water withdrawals and return flows by users of the different water sectors. The main objective of this research is to provide a sufficiently refined and consistent set of observations, both in space and time, with which to examine the salinity dynamics of the deltaic SAR system. The methodology used in this study was based on employing and combining various modelling approaches, underpinned by field data collected through a network of water quality sensors (divers). A systematic, comprehensive, and accurate monitoring program of salinity and water level over the entire length of the estuary and river (200 km) was developed for the first time under extremely harsh climatic and strict security conditions, yielding a unique dataset. Ten diver stations were installed at carefully selected locations, well maintained and regularly calibrated. Hourly observations of water level, temperature, and salinity variations during the full year 2014 were made. This part of the study determined the statistical, temporal and spatial distribution of salinity and its major causes, combining the 2014 data set with historical datasets gathered from local water authorities.

A quantitative analysis of the 2014 data set, combined with historical data, resulted in a broad description of the SAR's current state of hydrology and geography and the severe decline in water quantity and escalating levels of salinity over time (Chapter 3). The analyses covered the SAR as well as all contributing rivers (Euphrates, Tigris, Karkheh and Karun) with their connecting marshlands, which is essential to present a holistic picture. The analyses were based on the most recent data, though limited, on water availability, water resources development and management infrastructure, and water quality status. Water inflows were shown to have significantly reduced. The water quality status has deteriorated and by 2014 had reached alarmingly high levels, especially from Basra to the river mouth. The causes that could explain the steadily increasing water salinity varied from location to another. These include: decreased water quantity and quality (from the main and subsidiary water sources); seawater intrusion under tidal influences; poorly regulated localized water withdrawals; polluted return flows (from irrigation and several other wastewater discharge points); high evaporation rates; and occasional saline water discharges from the surrounding marshes.

Analysis of intra-annual variability of salinity levels shows high spatiotemporal variability in the range of 0.2-40.0 ppt [or g kg-1; ×1.5625 µS.cm^{-1}]. Similarities found in salinity dynamics were used to divide the river course into four distinct spatial units (R1-R4) to guide respective management actions (Chapter 4). Mean monthly salinity ranges of 1.0-2.0, 2.0-5.0, 1.0-12.0 and 8.0-31.0 ppt were observed for stretch R1 (Qurna to Shafi), R2 (Makel to Abu Flus), R3 (Sehan to Dweeb) and R4 (Faw near the estuary), respectively.

Correlating longitudinal and vertical salinity measurements provided the initial estimates of the extent of inland seawater excursion into the SAR estuary. To achieve a more physically based estimate of the seawater intrusion distance, a predictive model was developed that takes account of the specific tidal, seasonal and discharge variability and geometric characteristics of the SAR (Chapter 5). Seawater excursion was simulated analytically using a 1-D analytical salt intrusion model with recently updated equations for tidal mixing. The model was applied under different river conditions to analyse the seasonal variability of salinity distribution during wet and dry periods near spring and neap tides between March 2014 and January 2015. A good fit between computed and observed salinity distribution was obtained. Estimating water withdrawals along the estuary improved the performance of the model, especially at low flows and with a well-calibrated dispersion-excursion relationship of the updated equations. Seawater intrusion lengths, given the current measured data, varied from 38 to 65 km during the year of observation. At extremely low river discharge, a maximum distance of 92 km is predicted. These new predictions demonstrate that the SAR, already plagued with extreme salinity, is quickly approaching a situation where intervention will be either ineffective or much harder and costly. Several scenarios were subsequently investigated to demonstrate this point.

A 1-D hydrodynamic and salt intrusion numerical model was applied to simulate the complex salinity regime due to the combined effect of terrestrial and marine sources (Chapter 6). The model relied on the hourly time-series data for the year 2014. With the model, the impact of different management scenarios on the salinity variation under different conditions was analysed. The results show high correlation between seawater intrusion and river discharge. Increased uses of water upstream and increased local water withdrawals along the river will further contribute to seawater intrusion and increase salinity concentrations along the SAR. Improving the quantity and quality of the upstream water sources could reduce salinity concentrations. Discharging return flows from human uses, though saline, back into the river could counteract seawater intrusion, considering that the location of such outfalls affects both the salinity distribution and extent. The numerical scenario analysis based on SAR-calibrated parameters was particularly useful to study the longitudinal salinity variation under extreme conditions for any of the variables. With the assumed worst-case scenarios, best water management strategies can be screened but this requires a tradeoff analysis between water withdrawals and water salinity.

A multi-objective optimization-simulation model was developed for this (Chapter 7). The combined salinity system, including upstream salinity sources, return flows, and seawater intrusion, was simulated using a validated hydrodynamic model, which models salinity distribution in the river for different water allocation scenarios. Six scenarios were examined. The model was capable of determining the optimal solutions which minimize both river water salinity and the deficit of water supply for domestic use and irrigation. The model was used for exploring the trade-off between these two objectives. The developed approach combining a simulation and an optimization model can inform decision making for managing and mitigating salinization impacts in the region.

Results from the combined approach with simplified assumptions reduced a rather complex water system into a manageable 1-D model. The novel datasets and consequent analysis steps resulted in a new decision support tool which, with further refinements, can accommodate more complex scenarios. The study concluded (Chapter 8) that understanding of the prevalent high level of salinity variability in a complex and dynamic deltaic river system, a sound foundation of which has been laid by this study, does play a central role in designing measures to ensure the sustainable use and management of a water system. The comprehensive and detailed datasets formed the basis for a validated analytical model that can predict the extent of seawater relative to other salinity sources in an estuary, and to build a hydrodynamic model that can predict salinity changes in a heavily utilized and modified water system. The adaptability of the models to handle changing conditions makes them directly applicable by those responsible for water management. The procedure can be applied to other comparable systems.

The current efforts on salinity management are not enough to adequately address the mounting crisis. Continuous monitoring of water quality can localize and assess the relative impact of the various salinity sources at different times, particularly seawater and local sources of salinity. Managing seawater intrusion and local effects must take into account variations in quantity and quality of irrigation return flows and wastewater discharges along the SAR, as well as in the Euphrates, Tigris, Karkheh and Karun rivers. The crisis can only be averted through the cooperative water management initiatives taken by all the riparian countries, which require a paradigm shift from the current approach of unilateral water management planning to international cooperation and management on the shared water resources. Support from the regional and international community can contribute to this paradigm shift. The crisis mitigation strategies should find ways of increasing inflows from the upstream source rivers and improving their water quality. At the same time local measures are required to avoid drainage of poor quality domestic and industrial effluents and highly saline water from the marshes into the SAR. These efforts should be supported by sound scientific information.

x

TABLE OF CONTENTS

1 INTRODUCTION

1.1 Background

1.1.1 Water availability and water scarcity

Water covers about two-thirds of the earth's surface. Rivers and lakes form 0.3% of the world's freshwater and are considered the major water source for human use and consumption (Korzun et al., 1978; Karamouz et al., 2003).

Human population growth is the main driver of rising global demand for water and food products. The world's population is expected to increase to 9.2 billion by 2050 (see Figure 1-1). Rapid population growth, together with an anticipated increase in manufacturing and agricultural production, put more pressure on the environment, especially on the already strained water resources. The renewable freshwater availability was around 17,000 m^3 per capita per year in 1950s. In 2000, this was reduced to 7,000 m^3 per capita per year, indicating a decline of about 60 percent from 1950 to 2000. By the year 2050, water availability is expected to decline further to 5,000 m^3 per capita per year, which will be a decline of about 70 percent compared to the water availability in 1950s (see Figure 1-1).

The international average of available freshwater does not reflect the real distribution of available water across the globe. It does not show the regions with abundant water or the ones that experience scarcity. Uneven distribution of water resources exacerbates water scarcity. The spatial and temporal distribution of water availability varies in different regions. There is often too much or too little water. These fluctuations result in floods and droughts. Water is not always in the right place at the right time to meet the demands. Table 1-1 shows the regional distribution of available water resources of the world and the fraction of all available renewable water resources that are withdrawn and used.

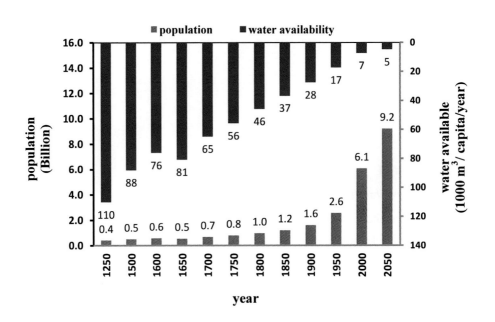

Figure 1-1. *World population and water availability (Data source: Dolphin, 2007; UN Water, 2012; UNDP, 2006) (based on 44,000 ×10^9 m^3/yr - the total volume of water transferred by the Earth's hydrological system to the land each year).*

By the year 2025, around 3 billion people could be living in water stress. And around 1.8 billion people will live in 14 water-scarce countries (Seckler *et al.*, 1998; Alcamo *et al.*, 2000; Vörösmarty *et al.*, 2000; UNDP, 2006). Countries in arid and semi-arid regions are the most vulnerable. Most of the countries in the MENA region (Middle East and North Africa) cannot meet their current countries or regions facing water demands (World Bank, 2006). Water security is of crucial importance for countries in the WANA region (West Asia and North Africa) (El-Kharraz *et al.*, 2012). The rapid socio-economic and technology changes in the Mediterranean region increase the environmental and water scarcity problems (Iglesias *et al.*, 2007). In these regions, 53 percent of the per capita annual withdrawals for all users are less than 1000 m^3 and 18 percent are below 2000 m^3 (Pereira *et al.*, 2002). According to the United Nations Development Program (UNDP, 2006), there are 43 countries with total population of 700 million suffering from water stress. The most water-stressed region in the world is the Middle East, while the largest number of water-stressed countries is in the Sub-Saharan Africa.

Table 1-1. *Availability of renewable water resources of the world (Data sources: Pereira et al., 2002).*

Country Group	Total annual internal renewable water resources 10^6 m^3/yr	Total annual water withdrawal 10^6 m^3/yr	Annual withdrawal as share of total water resources %	Annual internal renewable water resources per capita m^3/capita/yr	Percentage water withdrawal per sector		
					Agriculture %	Domestic %	Industry %
Sub-Saharan Africa	3713	55	1	7488	89	8	3
East-Asia and Pacific	7915	631	8	5009	86	6	8
South Asia	4895	569	12	4236	94	3	3
Europe	574	110	19	2865	45	13	42
Middle East and North Africa	276	202	73	1071	89	6	5
Latin America and the Caribbean	10579	173	2	24390	73	16	11

Water scarcity is a growing threat to humanity and environment. Four main drivers will increase water scarcity (UN Water and FAO, 2007). The first driver is the sufficient food production required for sustaining the population growth (Alcamo *et al.*, 1997; Shiklomanov, 1998; UNDP, 2006). The second is the expansion of existing urban areas and the creation of new cities. In 2010, urban areas were home to 3.5 billion people, which are expected to rise to 6.3 billion by 2050. Developing countries, where most projected urbanization growth will occur, are expected to double the population from 2.6 billion in 2010 to 5.2 billion in 2050 (United Nations, 2011; Varis *et al.*, 2012). Freshwater courses have limited capacity to respond to increased demand, and to process the pollutant charges of the effluents from expanding activities. Third, lifestyle and increased human development will increase per person domestic water requirements. Fourth, climate change will induce considerable changes on water resources. Freshwater availability will change at a regional scale, even with uncertain magnitudes. Arid and semi-arid regions will probably face frequent intensified drought periods as a result of increases in the variability of precipitation (Bates *et al.*, 2008). This will influence water availability of already stressed resources in the arid and semi-arid regions and will add more challenges to the management of water resources in the Middle East and North Africa, South and Central Asia, and parts of North and South America (World Bank, 2006; Iglesias *et al.*, 2007; El-Kharraz *et al.*, 2012).

While water quality is often a serious problem, salinity is believed to be largely affecting the lower reaches of rivers, especially in arid and semi-arid regions. The level of salinity in rivers has increased due to human activities such as polluted return flow from irrigation practices and reservoir evaporation. Hence, salinity changes are a major environmental problem facing many freshwater bodies (Huckelbridge *et al.*, 2010; Margoni and Psilovikos, 2010).

Extensive water use for various purposes, including building dams and canals to regulate river flows for various users, has altered natural flow systems and reduced the areas of downstream wetlands (Kingsford and Thomas, 1995; Lemly et al., 2000; Galbraith *et al.*, 2005; Jones *et al.*, 2008). Deterioration of water quality due to industrial effluent, urban pollution, and return flows from irrigation result in more water stress and public health issues, especially for the downstream users (Falkenmark *et al.*, 1999; World Bank, 2006; Iglesias *et al.*, 2007; El-Kharraz *et al.*, 2012).

Water salinity is a critical characteristic of estuaries and tidal rivers. Salinity is a major indicator of the water quality in estuarine systems (Jassby *et al.*, 1995; Reinert and Peterson, 2008). Tidal fluctuation, increasing freshwater diversions, and successive drought periods all contribute to more active salinity intrusion, increased inland excursion, longer duration, and stronger intensity (Jassby *et al.*, 1995; Xinfeng and Jiaquan, 2010). Salinity intrusion has serious effects on society, due to the need for freshwater for agricultural, industry, and water supply (Xinfeng and Jiaquan, 2010; Zhang *et al.*, 2011). Salinity can also constitute a serious problem to the physical and biotic components of aquatic ecosystems (Hart *et al.*, 1990; Roos and Pieterse, 1995; Nielsen *et al.*, 2003). The impacts of different factors on river water salinity changes will influence water availability of already stressed resources in arid and semi-arid regions. Figure 1-2 shows water scarcity distribution in the globe. Absence of sufficient institutional and financial resources will exacerbate the situation. Societies will not be able to maintain the health of these ecosystems and manage the available water resources sustainably without increasing the efficiency of infrastructure and developing capacity.

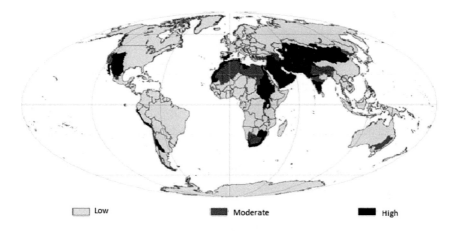

Low   Moderate   High

Figure 1-2. *Global Distribution of Physical Water Scarcity by Major River Basin (source: FAO, 2012a).*

1.1.2 Seawater intrusion

The application of any future management scheme to control water salinity must be closely evaluated prior to implementation. Therefore, the ability to predict the expected salinity changes within the river is necessary for any proposed control scheme. Numerical models are widely used to simulate the tidal current, salinity distribution, and transport processes in coastal and estuarine water systems. These models can be one or multi dimensional (e.g. Savenije, 1993; Ji *et al.*, 2001; Casulli and Zanolli, 2002; Chen and Liu, 2003).

Savenije (2005) provides a comprehensive theoretical background on estuary shape and a wide range of analytical equations that predict seawater intrusion and mixing processes in alluvial estuaries. Liu *et al.* (2004) studied seawater intrusion in the Tanshui River estuarine system (Taiwan) applying a 2-D numerical model. The study concluded that two effects, when combined, could cause salinity increases: (i) reduced freshwater flow due to the construction of two reservoirs and various other projects, which reduce the freshwater inflow, and (ii) the enlargement of the river canal due to riverbed degradation.

Liu *et al.* (2007) have established the relationship between freshwater flow and the distance of seawater intrusion. They applied a 3-D numerical hydrodynamic model to investigate the effect of water inflows on seawater intrusion in the Danshuei River estuary of Taiwan. The exponential relation is found to be a useful tool for predicting the distance of the seawater intrusion for any given discharge. Also, they found that the river discharge has a significant influence on salinity intrusion in the estuarine system. Simeonov *et al.* (2003) developed a multivariate, statistical approach to deal with spatial and temporal variations in water chemistry. This approach is useful for producing good information about the quality of surface water in cases dealing with large and complex databases. Complex interaction processes among flows, salt distribution, tidal currents and control structures characterize the estuary water system's behavior. For better-integrated water management, the proper model must be selected based on a system behavior investigation (Xu *et al.*, 2011).

1.1.3 Water Resources Management

Water resources management consists of plans and actions to manage the available water resources in an efficient and equitable manner. It should aim to satisfy all water users' requirements and control the impacts of flood, drought and water pollution (Koudstaal *et al.*, 1992; Loucks *et al.*, 2000). Water resources management comprises the totality of tasks to utilize water resources in an integrated manner, which contributes to socio-economic development (Koudstaal *et al.*, 1992). Water management activities allocate water as function of various economic, environmental, ecological, social, and physical objectives. It also involves all relevant disciplines and stakeholder decision-making processes for long term benefits (Loucks *et al.*, 2000; Bucknall, 2007). Understanding and monitoring water fluxes between upstream and downstream portions of a river basin is necessary for proper water resources governance (Van der Zaag, 2007).

1.1.4 Water Allocation

Excessive water abstraction and increased competition for available water resources could reduce water flows in rivers and hamper development in the downstream areas. Various simulation and optimization tools are used to support water allocation decisions (e.g. Loucks and van Beek, 2005). Weifeng and Chesheng (2010) applied a dynamic optimization model to improve the efficiency of water use. An increasing imbalance between the supply and demand may result in serious ecological problems. Therefore, Mckinney and Cai (1996) suggest a multi-objective optimization model for water allocation.

Application of optimization techniques to water resources allocation can provide optimal system management for different degrees of complexity. The optimization model solves water resources allocation problems based on an objective function and certain constraints considering different factors. The relation between these factors can be linear or non-linear. Usually, linear programming is used for water allocation models. The major challenge in using a linear program is the nonlinear behavior of certain important aspects such as evaporation and returns from demands. Haro *et al.* (2012) presented a generalized optimization model for solving water resources allocation issues in the Duero River basin in Spain. The model considered two non-linear constraints solved by an iterative methodology.

To overcome the negative effects and avoid the complex computations using many criteria in the classical water resources optimization methods, Gong *et al.* (2005) applied the rough set approach for water resources allocation in arid regions. Usually, the long-term, mean water availability provides the baseline for water resources management in a river basin. Shao *et al.* (2009) adopted the concept of water use flexibility limit to water shortage in the Yellow River Basin of China. This method was used to design a future water resources allocation system that guarantees efficient water allocation among different water users during climate-induced, inter-annual water variability. Water allocation among conflictive objectives is not easy task and required comprehensive evaluation for the present condition and analysis of possible management actions. The task will be more complex considering uncertainty of both water quantity and water quality.

1.1.5 Environmental Flow and Water Quality

Environmental flows are considered an essential issue in sustainable water resources management. Maintaining the environmental flow requirement is necessary for maintaining the riverine ecosystem health (*King et al.,* 2003). Yang *et al.* (2009) assess how much water should be left for river, wetland and estuary ecosystems in the Yellow River Basin (China). Based on the natural and artificial water consumption in the river basin, they determine the environmental flow requirements. They also take into consideration the classification and regionalization of the river system and multiple ecological objectives. Meijer *et al.* (2012) present a new functionality concept, which more realistically includes environmental flow requirements in water planning models. This concept helps avoid unnecessary environmental water allocation that can result from representing the environmental flow requirement as a fixed discharge per time step.

Instead of considering water quality as a constraint inside a water management decision support system, Paredes-Arquiola *et al.* (2010) developed a water quantity model and a water quality model for a two-fold water problem in the Jucar River Basin (Spain). The quantity problem was due to extensive of the agricultural water use, and the quality problem was due to point and non-point pollution sources. The whole process of water resources allocation in river basins can be demonstrated by using the integrated water quantity-quality method. In this method water quality is considered a key factor in water resources allocation (Wang and Peng, 2009).

1.1.6 The need to consider combined salinity sources

Increasing salinity (Total Dissolved Salt, TDS) is a major water quality problem in many rivers in the world, particularly in arid and semi-arid regions (Thomas and Jakeman, 1985; Shiati, 1991; Roos and Pieterse, 1995). At salinity levels greater than 1 ppt (part per thousand or kg/m^3) water becomes undrinkable (WHO, 1996). Above 3 ppt, water is no longer suitable for most agricultural uses. Irrigation with high saline water causes yield reduction proportional to the crop tolerance to the salinity (FAO, 1985; Rahi and Halihan, 2010).

Several studies have examined the discharge-salinity intrusion relationships (Zhichang *et al.*, 2001; Nguyen and Savenije, 2006; Nguyen *et al.*, 2008; Xue *et al.*, 2009; Whitney, 2010; Becker *et al.*, 2010). For example, *Wang et al.,* (2011) investigated the alteration of plume dynamics due to the abrupt increase or decrease in river discharges. Studies have also reviewed the impact of freshwater inflow on the salinity excursion (Bobba, 2002; Liu *et al.*, 2004; Vaz *et al.*, 2005; Liu *et al.*, 2007; Vaz *et al.*, 2009), In Das *et al.* (2012) different discharges scenarios of the lower Mississippi River were selected to control the salinity in the Barataria estuary. The results indicated that freshwater discharges strongly affect salinities only in the middle section of the estuary. Another set of studies assessed the influence of river discharge management on the salinity of estuaries (Myakisheva, 1996; Fernandez-Delgado *et al.*, 2007). Chen *et al.* (2000) concluded that the time lag of salinity in the river depends on whether the upstream freshwater inflow are increasing or decreasing and the magnitude of the flow. They also showed that the time lag for salinity is longer for decreasing inflows than increasing inflows, depending on the magnitude of the flow. Many investigations have explored the measures for salinity control of the river water mainly caused by irrigation practices (Thomas and Jakeman, 1985; Shiati, 1991; Roos and Pieterse, 1995; Kirchner *et al.*, 1997; Quinn, 2011). Xinfeng and Jiaquan (2010) showed that with increasing demand of water resources in both quantity and quality, the measures for preventing salinity intrusion should be proposed. Nevertheless, utilization of water resources should be ensured to support sustainable development and social stability.

However, little is known about salinity management when there is a combination of different salinity sources in a tidal river. There is lack of studies and thus understanding on the factors that determine the salinity of the tidal river, including irrigation practices, industrial effluents, urban discharges, quality and quantity of upstream river inflow, and seawater intrusion. The information on these factors provides the scientific basis needed to explore effective measures for controlling water quality and resources management. Moreover, excessive water diversions and practices upstream in the basin result in a reduction of water flow and deterioration of water quality in the lower reaches of rivers. This deterioration increases the competition of different stakeholders and sectors for available water resources. Ensuring minimum flows to control seawater intrusion, makes it appropriate to express the water management goals as being multi-objective rather than single objective.

Water resources optimization models have been used to determine optimal water allocations among competing water uses. Engineering optimization approaches have been applied as effective tools for planning purposes as well as real-time operation. They have also generated solutions for complex optimization problems (e.g. Windsor, 1973; Yeh, 1985; Barros et al., 2003; Labadie, 2004; Farthing et al., 2012; Singh, 2012; Kourakos and Mantoglou, 2013). Water resources optimization models offer the opportunity to perform sophisticated assessments of the natural, physical, and human-water system components that characterize the river basin. In this way, integrated hydrologic and economic models are well equipped to find optimal water allocation strategies in river basins (Mckinney et al., 1999; Cai et al., 2006; Mayer and Munoz-Hernandez, 2009). Mayer and Munoz-Hernandez (2009) overviewed the state-of-the-art of integrated water resources optimization models. Based on their review, they identify the need for improvements in inclusion of environmental flow; return flows from agricultural drainage, municipal and industrial wastewater. The studies indicate the need to advance water resource management optimization models to include the human and natural resources impacts of salinization due to wastewater returns.

Therefore, there is need to increase understanding of salinity dynamic associated with different factors. Take into account the impact of the return flows under tidal influences. This will provide the basis for evaluating different water management strategies to mitigate the impact of the salinity changes on the various water uses.

1.2   The study area: The Shatt al-Arab River

The Shatt al-Arab River (SAR) is formed by the confluence of the Euphrates, Tigris, Karkheh and Karun Rivers near the town of Qurna in the south of Iraq (Figure 1-3). The SAR is a 195 km long tidal river flowing south-eastwards, passing along the city of Basra, then the Iraqi port of Abu Flus and the Iranian port of Abadan, subsequently the city of Faw and from there, a final 18 km stretch where it discharges into the Arabian Gulf (also called the Persian Gulf, hereafter referred to as the Gulf). From Qurna to Basra there is a 63 km long stretch where the river elevation falls by about 0.7 meters. The river forms a part of the border between Iraq and Iran for the last 95 km of its course. Apart from the main tributaries, the SAR receives water from other rivers including the Garmat Ali, Ezz, and Sweeb. These rivers connect the SAR with the surrounding marshes.

The SAR width increases from 250 m downstream of the confluence to 700 m when entering the Basra region. The water depth varies from 6 to 13 m during dry periods (Marin Science Centre, 1991). The river is the most important water source for people in the region. It supports agricultural and industrial practices, navigation activities and ecosystem biodiversity. The water is diverted for irrigation purpose mainly for grain production in the upper course and palm forests in the lower course. Several water treatment plants divert water for domestic uses along the river. The rural communities, for whom agriculture and livestock is their main livelihood, use the water system for their activities and discharge their waste into it. The river water carries large amounts of silt coming from Karun. Siltation then necessitates continual dredging to allow navigation for medium and large fishing, transport, and vessels navigating to and from the Gulf.

1.3   Problem Statement

The Tigris-Euphrates river basin witnessed intensive development of water and land resources over the last 40 years. The riparian countries (Turkey, Syria, Iran, and Iraq) are engaged in construction of agricultural projects and hydropower plants. The largest effort is the ongoing South-Eastern Anatolia Project (GAP) in Turkey. It is the biggest multi-purpose water project in the basin, and one of the largest water resource

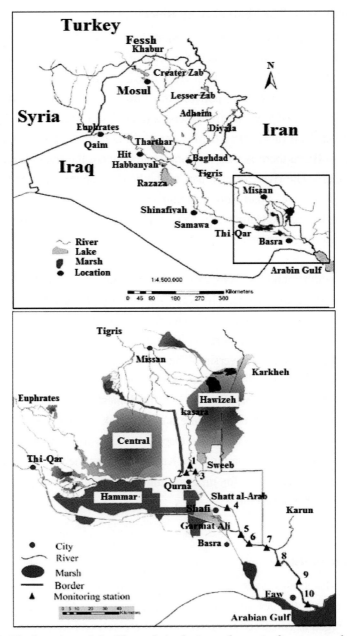

Figure 1-3. *The location of the Shatt al-Arab river, the main features and its tributaries, and measurement locations (numbered).*

development projects in the world. The project includes 22 dams and 19 hydropower plants. It was created to irrigate 17×10^3 km^2 of land and to produce around 7,526 MW of hydroelectric power. To date, 3×10^3 km^2 are irrigated as a result of this project (GAP, 2012).

The total projected demand of the riparian countries is higher than the amount of water supplied by the rivers. The total expected water demand after 2020 are 130 percent of the Tigris' water volume and 144 percent of the Euphrates water volume based on the highest demand scenario. On the Iranian side, the ongoing engineering projects are consuming and diverting water from all the tributaries draining into Tigris and Shatt al-Arab including the Karkheh and the Karun rivers (UN-ESCWA and BGR, 2013). The discharge of these rivers decreases to about zero during dry periods. As a result of increasing demands and allocations to various uses, there will be more tensions regarding equitable access between riparian countries, since water deficits are likely to undermine economic development as well as political stability in the area (Akanda et al., 2007; Jones et al., 2008).

Additionally, direct evaporation, from surface waters in rivers, lakes, reservoirs, and marshes, creates another problem: it leaves salt residues behind. The climate of the country is anticipated to become more arid to semiarid where the precipitation is much lower than evapotranspiration. In Iraq, the average temperatures range from higher than 35 °C in July to 10 °C in January. The mean annual rainfall is about 226 mm/yr, varying from 395 mm/yr in the north, to about 160 mm/yr in the middle and to about 125 mm/yr in the south (Figure 1-4). This sparse rainfall occurs during the winter season with most of the rainfall during January and March. The rest of the year represents the long dry and hot season without precipitation.

The Shatt al-Arab region has an arid climate with a hot, dry, long summer and short, mild winter. Large differences can be found in monthly temperatures and between day and night. The average temperature during the year is 25 °C. The average maximum temperature is about 31°C. The hottest month is July, with a monthly average of 43 °C, while the absolute maximum can reach as high as 50 °C during the hottest part of the day. The average minimum temperature is 19 °C. The coldest month is January, with the lowest average temperature ranging from 7.8 to

11°C (The Iraq Foundation, 2003; Italy-Iraq, 2006). An increase in temperature variability will aggravate the salt concentrations in freshwater due to expected increase in evaporation rates.

Various human activities throughout the countries sharing Tigris-Euphrates catchment areas have not only resulted in increasing water demand but also in decreasing water quality. Both rivers receive return flows, mostly without any treatment, these include huge amounts of agricultural drainage, municipal sewage, and industrial wastewater. The salinity of Euphrates at the Syrian-Iraqi border increased from 0.46 ppt in 1973 to about 1.3 ppt in 2009, after Turkey and Syria constructed a number of dams. The situation is better in Tigris at the Turkish-Iraqi border, where the majority of upstream water use is for hydropower generation. In this location, the salinity was about 0.24 ppt in 1973 and only increased to 0.3 ppt in 2009 (NCWRM, 2009; Al Amir, 2010).

At the downstream part, Iranian wastewater reaches the SAR through the shared marshlands (Hawizeh Marshes). In addition, the SAR receives sewage and industrial wastewater produced inside Basra province, Iraq. Several large factories, a paper mill, power stations, petrochemical industries, and refinery plants located along both sides of the river are known to discharge processed water. There are six creeks in the province and its surroundings are used as combined open sewage and storm drains that directly flow into the SAR.

The Shatt al-Arab region is now facing serious ecological problems and various challenges of water management. Water scarcity in terms of both water quantity and quality, particularly high salinity, is among the most pressing water management challenges. Climate variability and possibly climate change is the major reason for quantity variability, longer droughts, and variations in precipitation intensity. Previous upstream water diversion policies and ongoing water resources development strategies increased the impact of water shortages on the economic sector and domestic life in the region. There are extensive water withdrawals and diversions in the upstream, where there are 30 operational dams and 20 dams under construction or in planning (Al-Abaychi and Alkhaddar, 2010). The rapid establishment of new projects made competition obvious in response to the growing demand for freshwater.

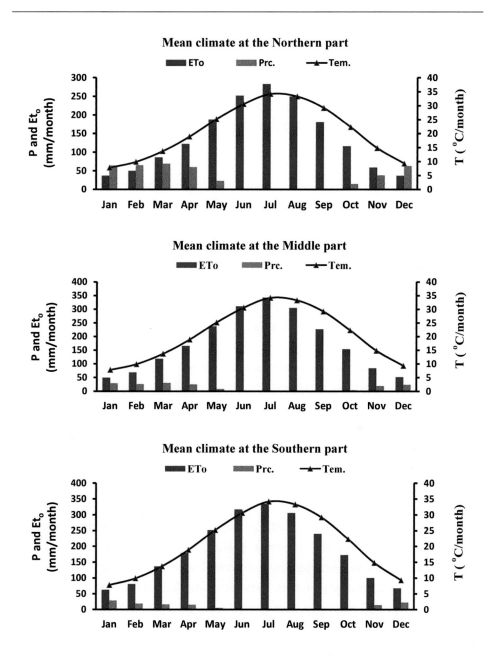

Figure 1-4. *Mean monthly precipitation (P), temperature (T), and evapotranspiration (ET_o) at the Northern (Mosul), Middle (Baghdad), and Southern (Basra) part of the country. (Data source: FAO, 2012b).*

Many of these projects are under development without adopting careful assessments of their environmental, hydrological, and socio-economic impacts on downstream areas. On the other hand, the inefficient irrigation practices and over-use of water by agricultural communities, domestic, and industrial sectors combined with their pollution effluents have brought drastic changes to the flow system and increased the salinity of the SAR.

Seawater intrusion is considered the major source of natural salinity in the Shatt al-Arab estuary. Seawater of the Gulf can reach up to considerable distances upstream during high tides. This distance increases with decreasing river flow. Freshwater-seawater interactions have never been scientifically studied for the SAR. Hence, there is a need to improve the knowledge about the relationship between tides and the river discharge over time. Better understanding of the tidal affect on the salinity regime of the river is essential to formulate better water management policies and strategies aiming at controlling rising salinity levels in the SAR.

There is limited domestic and international scientific literature available on Shatt al-Arab water management measures. The reason can be the very restricted data due to the majority of the basin being along an international border in an unstable area, which witnessed many prolonged wars. Isave and Mikhailova (2009) evaluated the morphologic structure and hydraulic regime of Shatt al-Arab mouth area, based on a review of studies on the Tigris and Euphrates runoff variation problem. Al-Abaychi and Alkhaddar (2010) evaluated the current situation of the SAR based on interviews with officials from the related ministries, and a three-day data collection excursion at three stations in the middle reach of the river (Basra upstream, Basra centre, and Basra downstream) including 15 water quality parameters. They identified high salinity concentrations resulting from different factors, including high hardness concentrations, and sewage pollution. They recommended a detailed study that covers a period of at least 12 months to understand the current issues and to suggest possible solutions.

The impact of upstream water practices along the Tigris, Euphrates, and other tributaries on the SAR have not been adequately evaluated, and there is a lack of understanding about their effects on river salinity. To date, the spatial-temporal trends of water salinity along the river have not yet been investigated in detail. As a result of all the above, poor water management activities are adopted that are not based on scientifically sound and reliable knowledge due to many reasons including the poor data availability. Therefore, there is need for a monitoring network to completely understand the impact of different sources of salinity and how the salinity changes with respect to various inflows through an entire hydrological year.

The waters of the Tigris and the Euphrates, also of the Karkheh and the Karun, are degraded during their journey from north to south. Water salinity reaches higher levels at the confluence, which is the start of the SAR. Spatial and temporal trends showing deterioration in water quantity-quality are obvious as a result of increasing consumption rates among the riparian countries, increased return flows, and intrusion of seawater. Salinity is expected to dramatically increase due to the inevitable rise of water consumption by different sectors, especially irrigated agricultural combined with increased human activities and this will aggravate the seawater intrusion. Figure 1-5 shows the schematic representation of the problem statement.

Salinity in the SAR is a concern because the salinity levels are above the limits for both drinking and irrigation water. Salinity is also highly dynamic, depending on the season and freshwater input. Basra province is facing a serious problem: the low quality of safe water for domestic and economic consumption. In most cases, the water of the public network is not fit for domestic use and is mainly used for cleaning. Drinking water is usually purchased from mobile water tanks or the markets (bottled water), which depends on the desalination treatment methods. Consequently, the salinity results in serious economic losses and has negatively affected industrial sectors and more investments are required to provide alternative drinking water.

Degradation of arable soils and water productivity results in the demise of agricultural lands and livestock. Increased salt content in the soil has severely damaged the date palm plantation industry. During the seventies the number of date palms was over 18 million, one fifth of the 90 million date palms in the world. By 2002 more than 14 million date palms were destroyed as a result of past wars and systematic destruction, increased salt content of soil and water, decreased flow, desertification, weathering, and pests. These impacts left around 4 million surviving

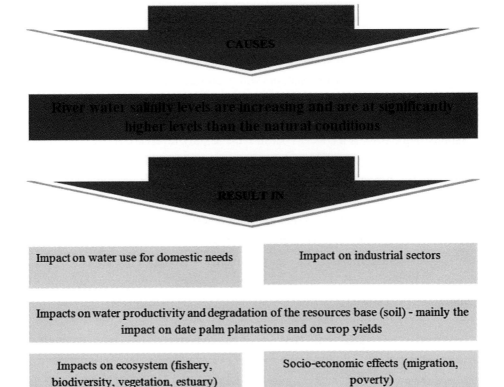

Figure 1-5. *The factors affecting salinity changes of Shatt al-Arab River and the main impacts on human and natural assets.*

trees in Iraq and Iran, with many remaining in poor conditions (Arabic Agricultural Statistics, 2002; FAO, 2013). Salinity has also had negative social effects in the region. The districts and sub-districts along the SAR live in the shadow of a continuing environmental disaster as animals died, trees perished and fisheries deteriorated, resulting in severe economic difficulties leading to a mass exodus towards Basra city. Whereas in the past fruit orchids were cultivated in the previously fertile agricultural areas, nowadays desert plants and wild tamarisk trees are found there. River water quality degradation also had potentially harmful impacts on regional fisheries. Salty water is not teeming with fish anymore and death of the wildlife has been accelerated (UNEP, 2001).

1.4   Research Approach

1.4.1 Research Objective

The literature review showed that numerous studies have examined the discharge-salinity intrusion relationship. However, there is a need to investigate the influence of multiple salinity sources on the salt balance in a tidal river. Further, little is known regarding how to simulate a complex ecosystem considering human-induced salinity and natural salinity. Water resources optimization models have been developed to produce appropriate water allocation strategies supporting sustainable water resources development. However, there is a need to assess the ability of water resources optimization models to account for the impact of return flows from agricultural drainage as well as municipal and industrial wastewater combined with the seawater intrusion on the human and natural resources.

With respect to the case study region, there is a need for further scientific assessment to explore the possible management alternatives for managing salinity dynamics. There is also a need to understand the current situation in the SAR and quantify the salinity changes that result from natural and human factors. The main objective of this research is to analyse water quantity and quality dynamics in a heavily modified tidal river and to study possible water and salinity management alternatives. This analysis would support water resources management efforts that are aimed at managing salinity levels with the purpose of avoiding or mitigating its negative impact on the human and natural resources.

1.4.2 Research Questions

The following research questions were formulated to achieve the main research objectives:

1- What are the main sources of salinity in a tidal river (Shatt al-Arab) and what are the relative weights or contributions of each?

2- What is the state of spatial-temporal variability of water salinity and water quantity across the river?

3- What is the impact of seawater intrusion on the river's salinity changes and how can we predict the seawater intrusion for any given inflow?

4- How can water and salinity dynamics resulting from different sources be best simulated using a physically based model for a regulated river with scarce data?

5- Can an optimization approach be reliably applied to develop water allocation strategies, which are designed to minimize the impact of the return flow and seawater intrusion on the salinity changes? Does such a model accurately predict water quantity-salinity relationship?

6- How can a combined use of models, including analytical, physically based simulation, and optimization models assist in informing decision-making processes?

1.4.3 Contribution of this study

This research deals with study of the salinity dynamics across a tidal river in arid region, the SAR, the lower reach of two main international rivers in the Middle East, the Tigris and Euphrates Rivers. The main contribution of this research is outlined below;

- Rather than studying the effects of a single salinity source individually, this study will consider possible salinity sources comprising man-made and natural sources simultaneously. The combined analysis and modelling of salinity sourced from the upstream catchment and from the sea side is one of the novel feature of this research.

- The research aims to extend the capability of the water resources optimization models to include the impact of return flow and sewer effluents in the optimization.
- Studying the salinity changes using an integrated approach underpinned by the integrated application of different tools, including analytical and numerical simulation and optimization models.
- This research will be the first study in the Shatt al-Arab region to provide alternatives for water management strategies that help to find suitable salinity levels along the river.
- This endeavour aims at consolidating the water resources management actions and plans, and provides options to mitigate the crises of water salinity affecting the society and environment in the region. The methods developed and findings of this research will also be instructive for other parts of the world with similar problems.

2   RESEARCH METHODS AND MATERIAL

2.1 Research process

The research process followed in this study is schematized in Figure 2-1. The research used several tools and methods, alone and in an integrated way, to understand the salinity variation along the river and to investigate the impact of different management actions. Application of these tools required intensive data to analysis hourly, daily and seasonal salinity changes. Consequently, they provide a better understanding and a baseline for water management interventions. The next sections briefly describe the methods used in this research.

2.2 Monitoring design and materials

The main function of any water resources model is to provide information about the system in order to support management planning and actions. The calibration of these models requires reliable and frequent water measurements. This information can be collected and generated by a monitoring network (Table 2-1). The monitoring objective is to provide significant information and data on water levels, salinity concentration, and the trends in water properties along the river considering the impact of tidal forces.

2.2.1 Water and salinity levels

Establishing the monitoring network requires designing the sample sites, variables, sampling frequencies and operational requirements. It is also based on the required data. The monitoring data will be mainly used to assess temporal and spatial trends of water salinity along the SAR.

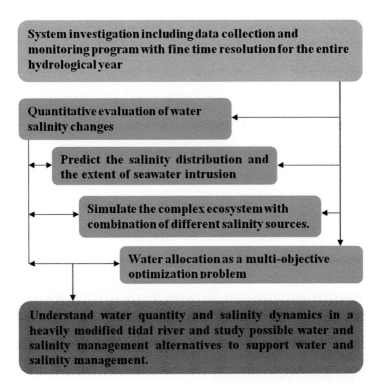

Figure 2-1. *The research process followed in this study.*

A monitoring network was established for this study with the main purpose of recording salinity of the SAR. Ten monitoring stations were installed along the river at important locations such as before and after the main confluence points and close to the Gulf to study the tidal influence (Figure 1-3). Figure 2-2 shows the schematic representation of the SAR and the location of the monitoring stations. The specific monitoring objective was to collect hourly data on salinity, water levels, and temperature. CTD-Divers (www.swstechnology.com) were installed to measure electrical conductivity EC (mS/cm), water temperature T (^{o}C) and water depth D (m). CTD-Divers are suitable for fresh and salt water application, and capable of storing around 48,000 measurements with registration frequency ranging from 1 second to 24 hours. This allows automatic recording and storing of hourly data for one year. The EC (mS/cm) was converted to TDS (ppt) based on an empirical relationship (1 mS/cm = 0.64 ppt) because TDS is the commonly used salinity indicator in the region. Most studies cited in this research also use this unit, which facilitates easy comparison.

Table 2-1. *The primary and secondary data used in this research with the required equipment.*

Category	Variable	Type	Details	Equipment
Hydrology	River discharge	secondary	Daily discharge at the upstream river measured for an entire year and discharge data collected from literature and Ministry of Water Resources (MoWR)	
	Water level	Primary	Hourly data at 10 stations for one year	CTD-Divers
Water quality	Salinity & temperature	Primary	Hourly data at 10 stations for one year	CTD-Divers
	Other parameter	Secondary	Monthly data collected at four stations for one year and water quality data at other stations from related ministries	
Climate	P, T, ET	Secondary	From the literature and related departments	
Water uses	Demand and abstractions	Secondary	From the literature and related departments	
Geometry	Cross-section	secondary	Use the available cross-sections from the MoWR and other departments	
Seawater Intrusion	Tidal Excursion	Primary	Following the tidal wave during spring and neap tide in wet and dry season, measure the salinity at different water depth along the longitudinal axis	Handheld depth-finder, GPS, speed boat

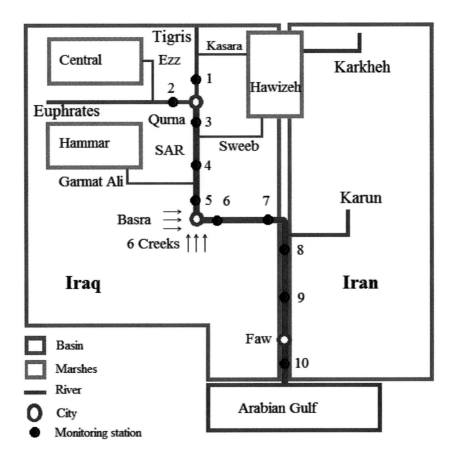

Figure 2-2. *Schematic representation of the main features of the SAR with the location of the monitoring stations.*

The monitoring stations were installed in places that ensured the divers were submerged during high and low tides. The other important considerations were the security of the divers, avoiding the navigation courses, and accessibility (see Table 2-2). Finding locations that fulfil these requirements was not easy, especially at the downstream portion where the tidal effect is most pronounced and the river is an international border with serious security issues. Authorization was granted to pass the several military points along the border, to get access to key locations and to safely conduct regular monitoring visits. Galvanized steel pipes were used to hold the divers in place (Figure 2-3). The pipes were provided with holes allowing the water to flow through. Depending upon the locations, the pipes were fixed (welded) to the columns of a water treatment plant jetty, platform of a harbour, or structure of a bridge. The divers required regular maintenance for cleaning the devices and tubes from sediments, water grasses and mainly barnacles (Figure 2-3) which can cause recording errors and may damage the divers. The top elevations of the tubes were previously set according to the local benchmarks based on the mean sea level using differential global position system (GPS). The data were collected with the support of the Water Resources Department (WRD) and the Marine Science Centre (MSC) in Basra during a complete year (2014), covering both the dry and wet season, low and high river flows and all tidal cycles.

Table 2-2. *The locations of the monitoring stations.*

Station	Name	Distance from the mouth (km)	Longitude	Latitude
S1	Tigris	197	47° 25.993' E	31° 01.853' N
S2	Euphrates	197	47° 24.583' E	30° 59.632' N
S3	Sweeb	190	47° 28.004' E	30° 59.741' N
S4	Shafi	169	47° 32.647' E	30° 50.876' N
S5	Makel	127	47° 46.768' E	30° 34.154' N
S6	Basra centre	117	47° 50.960' E	30° 30.630' N
S7	Abu Flus	97	48° 01.215' E	30° 27.542' N
S8	Sehan	69	48° 11.623' E	30° 19.587' N
S9	Dweeb	41	48° 24.067' E	30° 12.278' N
S10	Faw	10	48° 30.256' E	29° 57.879' N

2.2.2 Boat measurement method

One-dimensional analytical model was used to predict seawater intrusion along the Shatt al-Arab estuary. This model is based on a number of parameters which are collectable from field survey. Variables such as dispersion coefficient (D) and Van der Burgh coefficient (K) are not directly measurable and therefore they are obtained by calibrating the simulated salinity curve to the datasets form the sweater intrusion measurements. In this study, four measurement campaigns were conducted in the research area. The field surveys mainly focused on the measurement of salt concentration and water level, whereas data on the geometry and discharge were gathered from secondary sources. These campaigns were carried out for the entire hydrological year to cover different seasons during high and low flow. The measurements took place during the wet and dry periods at spring and neap tides: these were during 26 March 2014 (neap-wet), 16 May 2014 (spring-dry), 24 September 2014 (spring-dry), and 5 January 2015 (spring-wet).

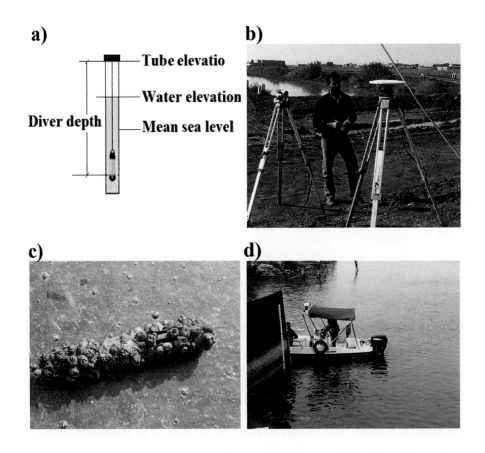

Figure 2-3. *Divers on the SAR: a) sketch of a diver installation, b) survey for installation of the stations, c) barnacles attached to a diver; and d) Example of a station at Makel (station S5).*

During a tidal cycle, the tidal velocity is near zero before the tidal current changed its direction. This situation is known as high water slack (HWS) right before the direction changed seaward, and low water slack (LWS) right before the direction changed landward. Salinity measurements were conducted at the moment before the flow changes its direction HWS and LWS. The HWS and LWS represent the envelope of the vertical salinity variation during tidal cycles, and are also used to determine the longitudinal tidal excursion. A moving boat technique was used in the field survey in which the boat moved with the speed of tidal wave to capture the slack moment (Figure 2-4). Starting from the mouth of the estuary and in the middle of the course, the salinity variations during tidal cycle were observed. A conductivity meter, YSI EC300A (https://www.ysi.com), with a cable length of 10 m was used to measure the vertical salinity profile for each meter depth from the bottom to the surface, and it was done repetitively for an interval of 3-4 km longitudinally until the river salinity is reached (in this case 1.5 ppt because the upstream salinity is also influenced by industrial and agriculture discharge).

2.2.3 River discharge and geometry

The required information on the river discharge and cross-sectional profiles was provided by the related local authorities. It is difficult to measure the discharge accurately in an estuary considering the tidal fluctuation. Hence, the discharge data from nearest (most downstream) station were used in the analysis. The daily stream flow data of the tributaries within the country were obtained from the Ministry of Water Resources (MoWR). However, there was no data on the discharge of the other main tributary, the Karun River, located in the neighbouring country, Iran. Experts in water resources authority, Basra indicated that the average discharge of the Karun estimated as 40 m^3/s. River cross-sectional data were collected based on the last survey carried out in 2012 by the GDSD (General Directorate of Study and Design) of MoWR.

Figure 2-4. *Boat measurement method; a) passing the military checkpoint and moving upstream keeping pace of the tidal wave during spring tide to the next measurement point, and b) water samples, conductivity meter, depth finder, GPS device and 10 meter cable.*

2.2.4 Water quality parameters

Other physical and chemical parameters also can be used to estimate the impact of human and natural factors on the water quality changes along the river, as presented in Table 2-3. The data collected from related local authorities including water resources, water treatment, and environmental departments, and covered two time periods, 1978 and 2014. These water quality indicators are selected considering the characteristics of pollution loads for different water uses as well as the existing spatial and temporal variations of the water quality variables.

Table 2-3. *Selected water quality parameters.*

	Parameter	Unit	Type	Description
1	PH		chemical	It measures the acidity or basic quality of water
2	Temperature	°C	physical	It is a controlling factor for the health of the river and can affect the other factors
3	Dissolved oxygen (DO)	%	chemical	It measures the amount of the gaseous oxygen which is important for aquatic life; it can be used as indicator for organic pollution
4	Na	mg/l	chemical	Sodium compounds are the most responsible for the salinity of the river and ocean.
5	Magnesium (Mg)	mg/l	chemical	It is the third most abundant element dissolved in seawater
6	Ca	mg/l	chemical	It is an important indicator determine of water utilization
7	Cl	mg/l	chemical	The major inorganic ion result from seawater intrusion and man-made activities
8	Sulphate (SO_4)	mg/l	chemical	It is a water quality variable that usually violates river water quality standards
9	Nitrate (NO_3)	mg/l	chemical	It is present in water contains animal and agricultural materials as well as domestic sewage; it can be used for point and non-point organic loads
10	Phosphate (PO_4)	mg/l	chemical	It is an important water quality parameter can be used to indicate water pollution
11	Turbidity	NTU	physical	It measures water cloudiness caused by suspended solids

2.3 Quantitative Evaluation

The statistical measures were calculated to quantify the changes of water quantity and water salinity. The quantitative evaluation was based on the information collected on salinity concentrations and water levels at each discharging source. For this the arithmetic means were computed for daily, monthly, and annual salinity. The median and quartile statistics were also determined for better understanding of the data distribution and the average conditions. The standard deviation was used to estimate the range of variability in the salinity data. The observed salinity data was analysed for presence of spatial and/or temporal correlation.

2.4 Simulation Model

To improve water management in the SAR, particularly in controlling the salt concentration to meet standards of drinking water for human and animal consumption and irrigation there is need to understand the impact of regional water development and various solution scenarios. This requires knowledge of the impact of water use, pollution sources, seawater intrusion, domestic and industrial return flows. Therefore, there is a need to use modelling tools, which help to get a better understanding of the interactions of different factors, and which allow to measure the potential impact of different strategies. This is to support decisions relating to water resource management. The main objective is to manage the salinity changes caused by combined sources, while it is relatively simple to model the impact of a single issue; it is much more complex to model multiple issues simultaneously. Two approaches have been used in this study to conduct the simulation of the system as briefly described below.

2.4.1 Analytical approach for estimating seawater intrusion

To avoid the complexity and heavy computational efforts of using multi dimensional models, an analytical model was used to simulate the longitudinal salinity distribution along the estuary caused by seawater intrusion. The analytical method is to understand the mechanism and behaviour of the model under any circumstances. Various authors have developed analytical approaches describing the salinity distribution in estuaries (Ippen and Harlemen, 1961; Prandle, 1985; Savenije, 1986). An analytical solution is able to provide important knowledge about the relationship between the tide, river flow, and geometric of the tidal channel. The one-dimensional modelling is usually based on a number of assumptions to simplify the set of equations. Several available models generally assumed a constant tidal channel cross-section to linearize and simplify the calculation processes. Here in this study the 1-D analytical seawater intrusion model proposed by Savenije (1986, 1989, and 1993) is considered as it requires the minimal amount of data. Another significance of this model is that it describes the geometry of an estuary in an exponential function. The model has been successfully applied to several single-channel estuaries worldwide (e.g. Risley *et al.*, 1993; Horrevoets *et al.*, 2004; and Gisen *et al.*, 2015a). Moreover, it can also describes the tidal propagation in multi-channel estuary (Zhang *et al.*, 2012) as well as estuaries with a slight sloping bottom (Nguyen and Savenije, 2006; Cai *et al.*, 2015). The model was used to evaluate the system performance in the real world under different configurations, and also to predict the extent of seawater in the upstream direction.

2.4.2 Physically based modelling for simulating combined salinity sources

Considering the complexity of the system, the application of numerical model was required to simulate water quality and quantity of the tributaries and upstream water flow, wastewater effluents, and water diversion for different water uses under tidal influence. The aim is to conduct the best simulation of the system considering an evaluation of different approaches, available data, and time.

For that purpose, a physically based distributed model, the Delft3D was used in this study to simulate the water system. Delft3D is an integrated modelling system for simulating the hydrodynamic and related process such as the transport of salinity (Lesser *et al.*, 2004). It has been developed by Deltares and is a flexible integrated program composed of several models, Delft3D-FLOW at the heart of it, which can simulate water flows, waves, water quality, morphology, sediment transport and salinity dynamics. The model can be used in natural and manmade environments, and can include density-driven flows. It solves non-linear differential equations of conservation of mass and momentum, under hydrostatic and non-hydrostatic free-surface flow conditions as described in detail in Lesser *et al.* (2004). These equations are derived from the three dimensional, Navier-Stokes equations for incompressible free surface flow. If the depth is assumed to be much smaller than the width, the vertical momentum equation is reduced to the hydrostatic pressure relation. Thus, vertical accelerations are assumed to be small compared to the gravitational acceleration and are therefore not taken into account. When this assumption is not valid then Delft3D provides an option to apply the so-called, Non-hydrostatic pressure model (Delft3D, 2012).

The 3-D Delft3D model was used to simulate the dynamic of the salt wedge and estuarine turbidity in the Rotterdam Waterway (De Nijs and Julie, 2012). Further, the Delft3D model and field observations were used to document the relationship between the salt wedge dynamics and the suspended particulate matter. The deterministic, 2-D hydrodynamic model (Delft3D) coupled with a heat transfer model was used to study the spatio-temporal pattern of water temperature dynamics in Arctic Sweden (Carrivick *et al.,* 2012). Delft3D has been used by several researchers to simulate tidal currents, sediment transport, and salinity changes in coastal areas (Ng *et al.*, 2013), lakes (El-Adawy *et al.*, 2013), estuaries (Van Breemen, 2008), and rivers (Van den Heuvel, 2010).

2.5 Integrated use of optimization with simulation modelling for water allocation

The multi-objective optimization approach was used for a water allocation process and to assess the impact of salinity changes on natural and human resources. The MATLAB software was used to write the optimization code and interlink the optimization and simulation models. The simulation model is used to determine the salinity changes correspondence for each set of decision variables, which are generated randomly by the optimization code. The results are evaluated by the objective functions for the purpose of optimization. The multi-objective optimization-simulation model was developed to identify efficient alternatives for water allocation that can serve the goal of improving the water management scheme. This was to minimize deficit of the water demands and to minimize the levels of river water salinity. The optimization model was investigated with different upstream conditions of river flow and salinity.

2.6 Performance evaluation of the model

A quantitative evaluation was conducted to compare the predicted and measured values. The model performance was evaluated by using three performance measures. Firstly, the root mean squared error (*RMSE*) was used to define the accuracy of the model. Secondly, another statistical index of agreement, the Coefficient of Determination (R^2), was used to measure the correlation between the simulated and observed water levels. Thirdly, the Nash-Sutcliffe Efficiency (*NSE*) was used to evaluate the prediction performance. The equations for computing these measures are given below:

$$RMSE = \sqrt{\frac{1}{n} \sum_{i=1}^{n} (P_i - O_i)^2} \qquad (2.1)$$

$$R^2 = \frac{(\sum_{i=1}^{N} (O_i - \bar{O})(P_i - \bar{P}))^2}{\sum_{i=1}^{N} (O_i - \bar{O})^2 \sum_{i=1}^{N} (P_i - \bar{P})^2} \qquad (2.2)$$

$$NSE = 1 - \frac{\sum_{i=1}^{N} (O_i - P_i)^2}{\sum_{i=1}^{N} (O_i - \bar{O})^2} \qquad (2.3)$$

where P and O are the simulated and observed variables and $\bar{O}$ and $\bar{P}$ are the simulated and observed means, respectively. The observed and simulated salinity will have the same units, ppt.

2.7 Cooperation with local authorities

This research is mainly based on the collection of detailed data which underpins the achievement of the study objectives. The data collection methodology included the monitoring network which provided information of salt concentrations and water levels in different sections along the SAR at an hourly time step. Ten locations have been chosen covering the sources which are expected to affect both the quantity and quality of the river water. The cooperation with local authorities was a key element of this whole research including field monitoring campaign and providing the required secondary data. The field work was carried out in coordination with the local university and water resources authority. The final accurate locations of the monitoring stations were determined based on the survey together with the delegates of Basra water resources department and marine science centre (MSC), University of Basra.

The monitoring campaign covered an entire hydrological year to provide data for the purpose of this research (including development of the models). The monitoring stations continued recording under joint supervision to provide a long-term and an appropriate database that can support the further future research concerning water quality and quantity in the region. The data is shared with the local institutions and will be a valuable addition to the ongoing efforts on monitoring of salinity in the study area.

3   WATER SYSTEM EVALUATION *

3.1 Introduction

Water limited environments covered about half of the global land area. These are mainly in arid and semi-arid regions, where the water scarcity has its roots (Parsons and Abrahams, 1994). Population growth is highly relevant to water sector problems and it is growing fast in many areas, particularly in the developing countries (Falkenmark *et al.*, 1999; Pereira *et al.*, 2002; Varis *et al.*, 2012). The likelihood impacts of climate change on rainfall patterns, in time and space will influence water availability of already stressed resources in arid and semi-arid regions (Bates *et al.*, 2008).

The SAR water system, in terms of both water quantity and quality, is under escalating pressure. The demands for freshwater have increased rapidly mainly due to growing population. Increases water allocation for different human activities not only have placed pressure on available freshwater, but also have led to water quality deterioration. The salinity have increased and caused serious repercussions on the environment and socio-economic sectors (Al Mudaffar-Fawzi and Mahdi, 2014).

A study by Isave and Mikhailova (2009) shows that flow regulation and water use for irrigation from the Tigris and Euphrates rivers by Turkey, Syria and Iraq have reduced the water and sediment inflow to the SAR. These changes have resulted in the degradation of the soil and vegetation and have enhanced the tidal influence. Although the study of Isave and Mikhailova (2009) provides a first comprehensive

---

* This chapter is based on Abdullah, D.A., Masih I., Van der Zaag P., Karim U.F.A, Popescu I., and Al Suhail Q., 2015. The Shatt al Arab System under Escalating Pressure: a preliminary exploration of the issues and options for mitigation. *International Journal of River Basin Management*, 13 (2): 215–227 [doi:10.1080/15715124.2015.1007870]

estimate of changes in the flows of the Tigris and Euphrates Rivers over the period 1928 to 1998, the study limits itself to the morphologic structure and hydraulic regime of the Shatt al-Arab mouth area. Information on river discharge does not cover recent conditions no planned infrastructure and water allocations or an analysis of water quality including salinity issues.

The UN-ESCWA and BGR (2013) inventory of the shared water resources in Western Asia noted the significant reduction of water inflows to the SAR owing to upstream water developments. The severe increase in salinity levels they reported is based on very limited data. The salinity increases were attributed to large scale water development projects upstream and to the drainage of the Mesopotamian marshes. While acknowledging the absence of international agreements and the presence of historic tensions and conflicts on the issue of international border along the SAR, this study emphasized the need of developing a joint water management strategy involving all riparian countries.

Despite the mounting environmental and socio-economic crisis in the region, there are very few local and international studies on SAR system which can provide the needed scientific analysis and potential options to avert the crisis. For instance, the local and international community took good notice of the deteriorating situation of the Mesopotamian Marshlands, which resulted in detailed scientific studies and practical efforts to avert the crisis (e.g. UNEP, 2001; Jones et al., 2008). Such efforts are however lacking for the Shatt al-Arab region. The lack of efforts could be attributed to the highly restricted data collection possibilities due to the fact that most of the SAR includes an international border between Iraq and Iran in an unstable region that has experienced several prolonged wars in the recent past.

A comprehensive analysis of the declining water quantity and quality, under current and future conditions, can provide a sound scientific basis for addressing these problems. This chapter analyses changes in water flows into the SAR under current and future conditions taking developments in the respective river basins in Turkey, Syria, Iran and Iraq into account. Changes in salinity levels and other quality parameters over time were studied and options for sustainable water resources management are highlighted.

3.2 The SAR water system

The SAR forms the downstream part of a system of rivers that feed it. The major upstream tributaries include the Euphrates and Tigris rivers that originate in Turkey, and the Karkheh and Karun rivers that are located in Iran. Important to note is that the first three tributaries flow through a complex system of extensive marshlands before they discharge into the SAR. This system of marshes is generally known as the Mesopotamian Marshlands. Each of these is briefly described in this section.

3.2.1 The Euphrates River

The Euphrates (Furat in Arabic) is the longest river in south-western Asia. Table 3-1 provides the main hydrological and geographical characteristics of the river. Its headwaters are the Murat and the Karasu Rivers in the highland of Anatolia, Turkey, which join at Keban and flow southward to the Syrian plateau, where the cultivable floodplain is no more than a few miles wide. In Syria, the river is joined by three tributaries: Sajur, Balikh and Khabur (Kolars, 1994; Shapland, 1997; UN-ESCWA and BGR, 2013). With a wide and slow stream, the river crosses the Syrian-Iraqi border at Qaim. Downstream, the river flows through the Jazirah plateau, a limestone desert. The river then loses its water into a series of braided channels, some of which flow into the Hammar marsh (Grego et al., 2004). However, it also receives water from the Tigris River through a canal from the Tharthar Lake constructed to supplement the decreasing flows in the Euphrates due to upstream developments. Then the river continues further through depression areas to finally unite with the Tigris River at Qurna, forming the SAR.

3.2.2 The Tigris River

The Tigris River (Dijla in Arabic) is the second largest river in south-western Asia. It has its springs in the Taurus Mountains in Turkey, about 30 km from the upper catchment of the Euphrates. The tributaries Batman, Garzan and Bokhtanchai contribute most of the Tigris' water in the upper reach (Italy-Iraq, 2006) (Table 3-1). The river flows southeast and forms the Turkish-Syrian border (approximately 37 km) and the Syrian-Iraqi border (approximately 7 km) (Kliot, 1994). The river flows toward the Mesopotamia plateau, and it is joined by five tributaries that contribute significantly to the river discharge, namely the Feesh Khabur, Great Zab (shared between Turkey and Iraq), Lesser Zab, Diyala (shared between Iran and Iraq) and Adhaim (wholly located inside Iraq). In Mesopotamia the Tigris waters are used for irrigation. Further downstream, at Missan, the river divides into five main branches and a series of canals. Four of the main branches eventually lose their water into the marshes, while the principal channel travels further south, joining the Euphrates at Qurna. Part of the marsh's water returns back to the Tigris through a number of small canals and mainly the Kasara River, north of Qurna. The river has other tributaries that are extremely downstream. All these tributaries join the mainstream on the left bank, and are detailed below, ordered from upstream to downstream (Grego *et al.,* 2004; Italy-Iraq, 2006; UN-ESCWA and BGR, 2013):

- The Feesh Khabur River, with a 6,143 km^2 catchment area, emerges from Sirnak, in eastern Anatolia. It runs for approximately 181 km south through Turkey and the north of Iraq, then spans west through Zakho. From Zakho, the river unit with the main tributary the Hezil Suyu, which forms 20 km of the Iraqi-Turkish border, then flow into the Tigris at the tripoint of the three countries (Turky, Syria, Iraq). The mean annual flow is 2×10^9 m^3 at Zakho, while it generates more at the confluence point with the Tigris. 43% of the total drainage area is in Iraq. The river or its tributaries are not regulated; however, it is mainly used for irrigation;
- The Greater Zab River (Upper Zab), the largest one, provides with the Lesser Zab the largest floodwater of the Tigris. It originates in Turkey and joins the Tigris 49 km downstream of Mosul. It provides the Tigris with an average flow of 13.18 $\times 10^9$ m^3 per year at the confluence point. The total river length is 462 km, with a catchment area of 26,310 km^2 of which 65% is in Iraq. Both riparian countries are planning to develop the water resource in the region. Turkey has three planned dams producing energy. Iraq has two dams: Bakhma dam is under construction and Mandawa dam is under planning designed for flood control, irrigation and hydropower generation;
- The Lesser Zab River (Lower Zab or Little Zab) emerges from the north-eastern Zagros ridge in Iran at an altitude of 3000 m asl. It is fed by snowmelt from Iran, runs through deep valleys and joins the Tigris 220 km north of Baghdad. The river has a total length of 302 km and a drain basin area of 19,780 km^2, of which 76% is

in Iraq and the remainder in Iran. The basin is equipped by two dams in Iraq territory, while Iran has one under planning phase. In Iraq, the Dukan Dam, at the upper part, is designed for flow regulation, irrigation and power generation, while the Dibis Dam, 130 km before joining the Tigris, regulates the discharge to a large irrigation project. The stream supplies the Tigris with $7.8 \times 10^9 m^3$ average annual flow at Altun Kupri station between the two dams;

- The Adhaim River originates in Iraq, formed by the confluence of two rivers at an altitude ranging from 1400 to 1800 m asl. It crosses the Hemrin Mountains and runs about 230 km to join the Tigris near Balad 80 km north of Baghdad. The river covers a basin area estimated to be 13,000 km^2 entirely in Iraq. It is an intermittent river, generates about 0.79×10^9 m^3 and is controlled by a multi-purpose embankment dam;

- The Diyala River has its source in the Zagros Mountains of Iran and is characterized by damming more than any other Tigris tributary and has many small shared tributaries. The Sirwan River is the largest of its tributaries, which is regulated by two multipurpose dams in Iran and another dam when entering Iraq. From that point, the Diyala River runs for about 130 km to meet its another main tributary the Wand River, where the Hemrin Lake is created by the Hemrin dam. The Wand also rises in Iran; a recent high exploitation of water resources in Iran territory resulted in an 80 % reduction of the Hemrin Lake capacity in Iraq. Diyala Wire, designed for irrigation purposes, is located 10 km downstream from the Hemrin dam. The river traverses a distance of 574 km to unite with the Tigris 15 km south of Baghdad. It drains around 33,240 km^2, with Iraq occupying a 75% share of the total area and Iran 25%, providing an average annual water volume of 5.74×10^9 m^3 at Baghdad;

- The Tib (or Teeb), Dewarege and Shehabi rivers originate in Iran and then flow through barren region in Iraq. They are mainly fed by rain water and have a high discharge in the period of intensive rains and experience scarcely extended drought periods in summer. They drain together more than 8,000 km^2 and bring about 1 $\times 10^9$ m^3 of highly saline waters, most of which empties in the marshes.

Table 3-1. *The length and drainage area of the Tigris and Euphrates Rivers in each country and water contribution of the riparians to the total flows.*

	Unit	Euphrates				Tigris				
		Turkey	Syria	Iraq	sum	Turkey	Iraq	Iran	sum	Total
River length	10^3 m	1,105	675	1,160	2,940	437	1,425		1,862	4,802
	%	38	23	39	100	23	77		100	
Catchment area	10^6 m^2	124,320	96,800	155,400	376,520	54,145	142,308	127,500	323,953	700,473
	%	33	26	41	100	17	44	39	100	
		At border	In Deir al Zor	At Hit		At border	In Baghdad	At border		
Mean annual runoff	10^9 m^3/yr	28.5	32.4	25.6	32.4	20.3	47.3	6.2	47.3	79.7
	%	88	10	2	100	43	44	13	100	
Contribution of the total annual flow	%	35.8	4.1	0.8	40.7	25.5	26.1	7.7	59.3	100

Data sources: Kliot, 1994; Kolars,1994 ; Shapland, 1997; UNEP, 2001; KHRP, 2002; The Iraq Foundation, 2003; Biedler, 2004; Altinbilek, 2004; Italy-Iraq, 2006; Geopolicity, 2010; UN-ESCWA and BGR, 2013; and authors' calculation.

Note: these figures exclude the contributions of the Karkheh and the Karun rivers.

3.2.3 The Karkheh River

The Karkheh River originates in the Zagros Mountains of Iran and drains an area of 51,000 km². It has total length of about 964 km. Its discharge flows into the transboundary Hawizeh marshes that straddle the Iraqi and Iranian border (Masih, 2011; Hessari et al., 2012). The total annual renewable water resource (both surface and groundwater) is estimated at 8.6×10^9 m³/yr of which 2.5×10^9 m³/yr was the total consumptive use in 1993-94. The major river regulation started in 2001 with the construction of the large and multipurpose Karkheh dam. The current water allocations are estimated around 4.9×10^9 m³/yr and expected to increase to 8.6×10^9 m³/yr in 2016, due to investment in various dams and irrigation schemes that are currently under construction or in the planning phase (Italy-Iraq, 2006; Masih, 2011). The Karkheh River used to contribute about 6.3×10^9 m³/yr to the Hawizeh marsh under natural flow conditions (Grego et al., 2004). However, the water flow to the marshes has been altered after the construction of the dams, and the discharge may be as low as zero in future with the accomplishment of upstream developments. Over the years, different water activities have negatively impacted its water quality, the salinity increased from 0.8 to 1.8 ppt during 1988-2002 (UN-ESCWA and BGR, 2013). Further details on the Karkheh River can be found in JAMAB (1999), Ahmad et al. (2009), Marjanizadeh et al. (2009) and Masih et al. (2009).

3.2.4 The Karun River

The Karun River is among the most important watercourses of Iran, with the largest discharge and being the only navigable waterway. It has total length of about 867 km. Similar to the Karkheh River it originates in the Zagros Mountains of western Iran. The river joins the SAR 87 km upstream of the mouth. The Karun River drains an area of about 63,000 km², which lies wholly in Iran (Italy-Iraq 2006). There is limited information about the Karun's contribution to the total volume flow into the SAR. Most relevant is Ahvaz station in Iran (UN-ESCWA and BGR, 2013; Salarijazi et al., 2012; Afkhami et al., 2007), the most downstream gauging station but still located approximately 200 km upstream of the confluence with the SAR. The Karun has an average discharge of 24.7×10^9 m³/yr (Grego et al., 2004). Due to large scale water developments, the mean annual discharge of the Karun has experienced a consistent negative trend from 818 m^3/s to 615 m^3/s before and after 1963, respectively (UN-ESCWA and BGR, 2013). Whereas Salarijazi *et al.* (2012) reported a mean annual river discharge at Ahvaz of 1,442 m^3/s for the period 1954-2005, the mean monthly river discharge for the period between 1978 and 2009 was less than half at 667 m^3/s. The amount of freshwater supplied by the Karun River to the SAR holds significant importance in terms of reducing seawater intrusion.

3.2.5 Mesopotamian Marshlands

Mesopotamian Marshlands (locally known as Al-Ahowar) is an extensive system of ecologically important wetlands formed at the confluence of the Tigris, Euphrates, and Karkheh Rivers. Al-Ahowar is composed of several lakes and permanent marshes that are connected and flow into each other frequently during high flow periods. They are surrounded by deserts and are generally divided into three areas, namely the Hammar, Central, and Hawizeh marshes (Figure 1-3).

Under natural conditions these wetlands once covered an area of approximately 15,000 to 20,000 km^2, and have been recognized as the greatest wetland region in the Middle East and Western Eurasia (UNEP, 2001). The marshes are primarily fed by flows of the Tigris, Euphrates and Karkheh rivers, which fork into series of braided canals that discharge into the marshes. The average depth ranges from 1 to 1.5 m and reaches up to 6 m in some places. The wetlands and surrounding areas are characterized by an extremely flat and very fertile alluvial plain, allowing extensive wheat and rice cultivation. The marshlands are home to around 500,000 people, locally named Marsh Arabs, who are descendants of the Sumerians. Most of the marshes are covered by natural plants, often reeds and papyrus, and are the habitat of several animals and fish and bird species. Its geographical location permits a stopover for some birds while migrating between Siberia and Africa. As such, the Mesopotamian Marshlands used to be an ecosystem and natural heritage of local and international importance that supported numerous species of wildlife and aquatic biodiversity (UNEP, 2001).

After the Gulf War in 1991, the Iraqi government intervened to reduce so-called evaporation losses occurring in the marshes. The Ezz (Glory) River and the Main

Outfall Drain (MOD) are both manmade canals constructed to collect water from the tributaries of the Tigris and Euphrates, respectively, routing them around the marshes and directing them into the Gulf. Due to the drainage of these marshes and the geopolitical situation many Marsh Arabs were forcibly displaced; around 40,000 people migrated to neighbouring Iran while up to 250,000 were internally displaced (UNEP, 2001). As a result the area was almost entirely dried out by 1995 and only minor scattered segments remain. Thus, the vast wetlands were destroyed and converted into desiccated desert lands.

Since 2003, national and international efforts have attempted to restore the ancient wetlands (e.g. Italy- Iraq, 2006; UNWP, 2011). However, the restoration of the desiccated marshes encounters many challenges. A further reduction of water flows into the marshes, especially after a series of recent projects created in the upstream regions, does not help. The marsh restoration endeavours involved collecting and analysing data, capacity building, developing remote sensing monitoring system, supporting related scientific studies (environmental, social, economic) and implementing basic infrastructures (water and wastewater treatment plants, health and education centres). Recently, the Marsh Arabs have begun to return after the marshes have partially recovered to some places not all the marshes.

3.3 Hydrology of the SAR

In this section the hydrology of the SAR is analysed based on limited data that have been published in the literature.

3.3.1 Inter-annual flow variation

It can be argued that the SAR flows from Qurna up to the confluence of the Karun mirror the combined effect of Tigris and Euphrates flows. The available records from the MoWR of the period 1933–2011 indicate that whereas both rivers show high intra- and inter-annual variability (Figures 3-1 and 3-2), their natural flow regimes were quite similar. Both rivers used to have similarly temporal discharge patterns with high and low flows recorded in the same years most of the time, and similar seasonality in flows. Floods and droughts and high intra- and inter-annual variability of flows have been major issues in the region, which triggered the construction of various hydraulic infrastructures in the past. The flow regimes of both rivers have dramatically changed over time due to massive infrastructure development for irrigated agriculture and hydropower (e.g. UN-ESCWA and BGR, 2013). Some of these points are substantiated below.

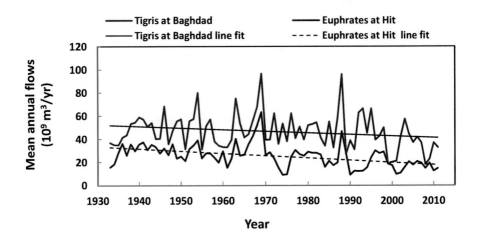

Figure 3-1. *The inter-annual flow variations of the Tigris and the Euphrates Rivers, 1933–2011, indicating a declining trend (0.14 × 10⁹ m³/yr for the Tigris and 0.19 × 10⁹ m³/yr for the Euphrates).*

Prior to the dams' construction, the Tigris and Euphrates rivers experienced a considerable variation in annual flows (Figure 3-1). The Euphrates is mainly fed by melting of snow in the highlands of eastern Turkey, as well as rainfall in the mountain areas in Turkey. The analysis of available data indicates that the average annual natural flow of the Euphrates at Hit in Iraq was 30.3×10^9 m³/yr during the 1933–1972 period. The river experiences a high annual variance in water volume, the lowest flow recorded at Hit being 15.2×10^9 m³/yr (in 1961) and the highest flow 63.3×10^9 m³/yr (in 1969). The average annual natural flow of the Euphrates at the most downstream station (measured at Thi-Qar gauging station) was 13.53×10^9

m^3/yr. The Tigris experiences a similar flow variation. During the 1933–1988 period, the lowest flow recorded at Mosul gauging station in Iraq was 11.5 × 109 m^3/yr (in 1973) and the highest flow was 43.2 × 10^9 m^3/yr (in 1969), with an average annual flow of 21.5 × 10^9 m^3/yr. The average annual natural flow of the Tigris at the most downstream station (measured at Missan gauging station) was 7.1 × 10^9 m^3/yr.

Figure 3-1 clearly shows the significant decreasing trend in the annual flows of both Euphrates and Tigris rivers over the period 1933–2011, which is attributed to water withdrawals for human uses. The sum of water flows from the two rivers is much less than Iraq's projected water requirements in 2015 of 67 × 10^{9}m^3/yr, as estimated by the MoWR and IEA (IEA, 2012).

3.3.2 Seasonal flows variation

The Tigris is characterized by high seasonal flow variation, as well as spatially along the river. The four main tributaries, whose sources are located in the Zagros Mountains, mainly affect the flow in the river, besides water diversions downstream (Figure 3-2). Most of the precipitation falls in winter and much of it falls as snow, with snow melt in spring coinciding with heavy rains in the Anatolian and Zagros Mountains, the Greater Zab and Lesser Zab, potentially doubling the volume of the Tigris during the high-flow season in March and April. Moreover, a large amount of water diverted from the Tigris to the Euphrates, mainly through the Tharthar depression in the middle stretch in Iraq and the Gharraf River in the lower stretch, has significantly altered the flow. The seasonal flows display a high variation in discharge. Peak flows come in April, averaging 1433 m^3/s at Baghdad, while the lowest come in September, when the average is 113 m^3/s (Kolars, 1994; KHRP, 2002).

The peak water flow is driven by winter rainstorms and spring snowmelt. The low flow extends from summer to fall and early winter. The high flows are extremely important to the SAR and the estuary, controlling the salinity and the extent of seawater intrusion from the Gulf. Upstream developments and water diversions obviously reduce the flow into the SAR (Figures 3-3 and 3-4).

The daily freshwater flows into the SAR have been greatly altered (Figure 3-5). These changes have affected the natural and human resources that depend on the river. The water resource developments in the watersheds of the upstream rivers capture a large proportion of the flow. This results in a significant reduction of the freshwater discharge, as illustrated by the flow conditions during selected years in Figure 3-5. The planned and ongoing developments aim to store peak flows that may still occur as in the hydrological year of 2002–2003 (Figure 3-5).

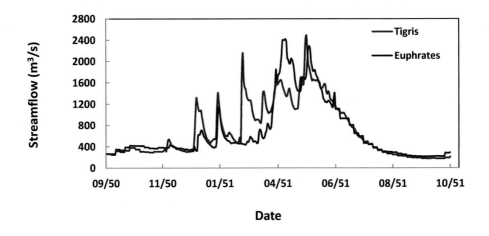

Figure 3-2. *Seasonal flow fluctuation of the Tigris and the Euphrates Rivers during the hydrological years 1950–1951.*

During recent years most of the SAR water is coming from the Tigris. The quality of the Euphrates's water has deteriorated as a result of accelerated water resources development upstream. Salinity of the Euphrates water often exceeds the allowable limits for drinking and irrigation, especially during summer. Then the Euphrates water is diverted to the marshes away from the SAR. The SAR receives water from the Euphrates only during high-flow periods when the water has a lower salinity concentration, normally in winter.

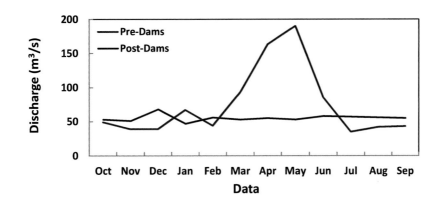

Figure 3-3. *The Tigris average monthly flow at Missan station of the hydrological years 1959–1960 (pre-dams) and 2000–2001(post dams).*

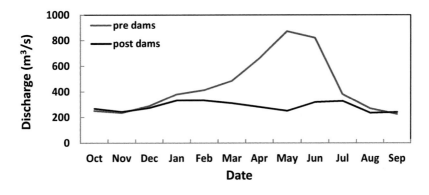

Figure 3-4. *The Euphrates average monthly flow at Thi-Qar station during the periods 1950–1980 (pre-dams) and 1982–1997 (post dams).*

3.4 Water quality

Data on water quality of the SAR are even more scanty than data on water quantity, which limits a proper description and analysis of water quality dynamics of this tidal river. The following sections briefly discuss the salinity status of upstream rivers based on available data as reported in the literature and found in unpublished reports, also discuss the variation of other water quality parameters over time and space along the SAR.

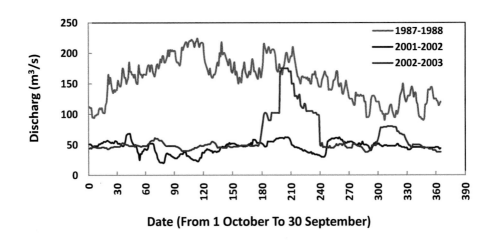

Figure 3-5. *The average daily discharge of the Tigris River upstream of the Shatt al-Arab (near Qurna city) before and after the development in the basin.*

3.4.1 Temporal and spatial variation in salinity

The salinity of the Tigris and the Euphrates waters is increasing in space, as the rivers flow downstream, as well as in time. Figures 3-6 and 3-7 represent the annual average salinity concentration based on data for several years from 1980 to 2002 as found from the literature and unpublished reports of different organizations. Various sources (The Iraq Foundation, 2003; Rahi and Halihan, 2010; UN-ESCWA and BGR, 2013) report that the salinity at Missan gauging station on the Tigris increased threefold in 22 yr (from 0.6 ppt in 1980 to 1 ppt in 1986 and to 1.8 ppt in 2002). This increasing trend is expected to continue, mainly driven by the decreasing flows combined with impacts of other human activities along the SAR. The salinity of the Euphrates at the discharging point into the SAR has also gradually increased over time and was more than 3.5 ppt in the year 2002 (Figure 3-7), which is twice the salinity of the Tigris' water. This explains why the MoWR blocks the Euphrates near the confluence point and diverts most of the water to the marshes to control the discharge of highly saline waters into the SAR. This alarming increase of the salinity concentration, which is likely to increase further, is caused by intensive water developments upstream of the SAR, i.e. a decrease in water flow and an increase in saline return flows of water uses such as irrigation.

The salinity of SAR has increased not only because of increased salt inputs from upstream into the SAR, including from the surrounding marshes, but also due to human activities along the SAR itself, as well as increased seawater intrusion from the Gulf occasioned by reduced inflows from the upstream rivers which allows the seawater wedge to move further upstream. Figure 3-8 shows the temporal salinity changes of the SAR near Basra city during 1980–2008.

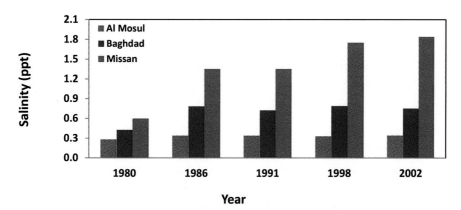

Figure 3-6. *Spatial and temporal salinity changes of the Tigris River, Iraq (Data sources: The Iraq Foundation (2003), Rahi and Halihan (2010), and UNESCWA and BGR (2013)).*

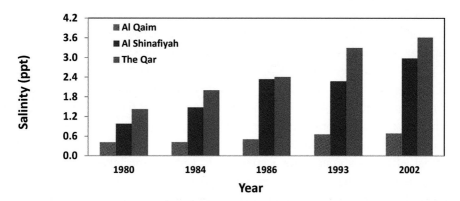

Figure 3-7. *Spatial and temporal salinity changes of the Euphrates River, Iraq (Data sources: The Iraq Foundation (2003), Rahi and Halihan (2010), and UN-ESCWA and BGR (2013)).*

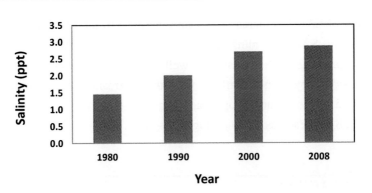

Figure 3-8. *The temporal salinity changes of the SAR at Basra, Iraq (Data sources: The Iraq Foundation (2003), Rahi and Halihan (2010), and UN-ESCWA and BGR (2013)).*

3.4.2 Water quality variability

   To evaluate the general trend of the SAR water quality variation, number of ions was selected (see Table 2-3). Mainly the ions washed out of the soil to river water, ending up in oceans. Also these ions end up in water from industrial and wastewater effluents in different compounds. Therefore, their concentrations depend on the geological conditions and effluent contamination. The water quality profile of the SAR, based on unpublished data collected from competent authorities, is presented in Figure 3-9, indicating mean annually values of 1978 and 2014. There are substantial variations in the chemical parameter levels over time and space. Generally, compared to the 1978, most of the chemical ions increased during 2014, mainly in the downstream direction, reaching the highest levels near the SAR estuary. Only nitrate (NO_3) level was higher during 1978, this could be related to higher drainage water from agricultural sector associated with extensive irrigation practices (Hussain, 2001). The sulphate (SO_4) level was higher during 1978 at the upstream of the river which reflect oil pollutions caused by heavy navigation of petroleum vessels along the river, while recently dominated at the estuary portion, this beside the other industrial effluents through the Karun River which contributed to increase the sulphate levels at the SAR estuary (Al Mahmod *et al.*, 2008). Increasing the concentration of most ions including Mg, Cl, Na, and Ca is attributed to high salinity levels (Moyle, 2014). Therefore, the highest levels of ions are recorded at the mouth of the river at Faw due to highest tidal influence combined with upstream saline inflows.

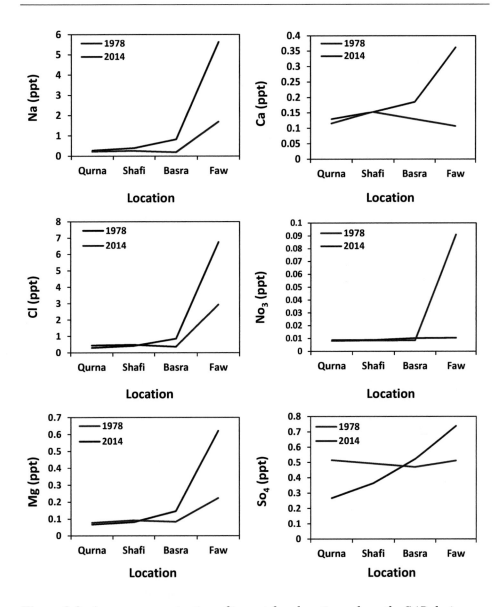

Figure 3-9. *Average concentration of ions at four locations along the SAR during 1978 and 2014, starting from the upstream of the SAR at Qurna and ending at the river mouth at Faw (see Figure 1-3 for those locations).*

Table 3-2 presents the annual average values of other physical and chemical parameters in range of locations in 2014. The pH is a measure of hydrogen concentration, range from 0 to 14, and determine of river water how acidic or basic. Considering the Iraqi requirements, the allowable limits fall between 6.5 and 8.5, where 7 represent the neutral condition. The measured pH along the SAR ranged from 7.96 to 8.14 (Table 3-2). The pH records vary slightly along the river around average of 8, indicating increasing alkalinity of the river water. This could be attributed to increase of the carbonate and bicarbonate concentrations resulted of high salinity levels (Al Sabah, 2007). Table 3-2 shows a small variation in concentration of the dissolved oxygen (DO), phosphate (PO_4), and water temperature (T) along the river. The amount of oxygen in the water was in the allowable limits (> 5 ppt). The records show a slight decline in dissolved oxygen level at the estuary, which could be due to discharging of organic materials from surrounding communities which consume more oxygen (Al Hajaj, 1997). The concentration of the phosphate is more correlated to the industrial effluents (Al Aisa, 2005; Jaafer, 2010).

Table 3-2. *Average concentration of quality parameter at four locations along the SAR during 2014 (see Figure 1-3 for those locations).*

	Qurna	Shafi	Basra	Faw
pH (mol/l)	8.14	8.14	7.96	8
DO (ppt)	8.05	8.13	8.1	7.25
PO_4 (ppt)	0.3	0.29	0.32	0.28
T (^{o} C)	22.5	22.8	23.4	23.4

The results presented in this section clearly show a significant deterioration of water quality over time and from upstream to downstream of the SAR. The major reasons are: decline in freshwater flows due to increased withdrawals and consumption; changes in the flow regime due to increased river regulations (dams, barrages and irrigation infrastructure); irrigation return flows having high levels of ions and organic materials; salt enrichment in reservoirs and marshes due to evaporation in particular in hot and arid environments; and disposal of untreated wastewater into the rivers. The salinity levels, represented by different compounds of ions are anticipated to increase in future, given the ongoing increased water withdrawals, pollution discharges and massive river regulation works in Turkey, Syria, Iran and Iraq. The current efforts on quality management, particularly the salinity, are not enough to adequately address the mounting crisis which is most alarming in the case of the SAR.

3.5 Water resources development and management

The name Mesopotamia evokes an image of fertile lands interspersed with rivers, which supported a number of ancient civilizations. Societies developed and expanded according to their ability of water control and irrigation management. For thousands of years, Mesopotamia has witnessed enormous irrigation and agricultural practices. The Sennacherib and Samara canals were the largest irrigation canals during the Assyrian period among 365 canals discovered in 1955 (Shayal, 2010). These canals were used to convey water from springs and fountains bound between highlands to the cultivation areas. In the sixth century, the Nahrawan canal represented the largest scale irrigation scheme in the region, with around 300 km of canal, providing irrigation water to the lands east of the Tigris River (Biedler, 2004).

3.5.1 Impact of water infrastructure

Large-scale water infrastructure developments accelerated in the twentieth century. Figure 3-10 shows the cumulative storage capacity of the major water resources development endeavours in the Euphrates, Tigris, Karkheh and Karun basins. All these projects are installed to serve the purpose of irrigation development, flood control and/or hydropower production. Iraq was the first country to initiate water resources development and flood control projects. However, currently, all basin countries (Turkey, Syria, Iran and Iraq) have developed large projects, most often unilaterally without consultation with the other riparians. The total storage capacity behind the existing dams is larger than the average annual flow of Euphrates, Tigris, Karkheh and Karun Rivers.

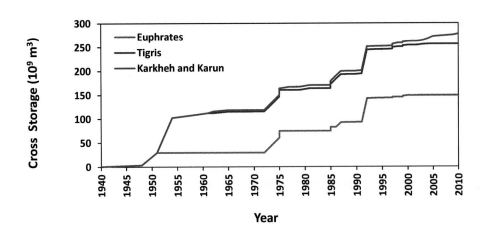

Figure 3-10. *Cumulative gross storage over time developed in the Euphrates, Tigris,*
and Karkheh and Karun Basin (Data sources: UNEP (2001), FAO
(2009), UN ESCWA-BGR (2013)).

Earlier studies have indicated that the demands for Euphrates water in Turkey and
Syria can be met, but with serious impacts for Iraq, which will be confronted with
significant water shortages (Kliot, 1994; Kolars, 1994). The situation on the Tigris is
better, since water availability is reported to be sufficient to meet the demands, with
some 'surplus' water left in the system. Despite large uncertainties in these
projections, one can infer that water shortages in Iraq will increase which will affect
all major water uses (e.g. agriculture, marshes, drinking, industry, environment and
ecology). Moreover, similar trends are anticipated as the result of current and ongoing
water resources developments in the Karkheh and Karun rivers in Iran, as well as the
tributaries of the Tigris River shared between Iran and Iraq. This point is illustrated
by Figure 3-11, which shows that the contribution of the Lesser Zab and Diyala, two
important tributaries of the Tigris that originate in the Zagros Mountains in Iran, to
the main Tigris has decreased consistently since 1990. Estimates of future water
demands under full development scenarios for the Euphrates and Tigris rivers, taking
into account expected population growth and large-scale development projects, show
that water flows to the SAR system will continue to decrease over time, which will
further deepen the ongoing water crisis in the region (Table 3-3).

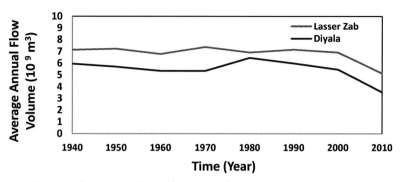

Figure 3-11. *Annual average flow alteration of the Lesser Zab and Diyala rivers.*

3.5.2 Impact of return flows

Return flows are an important element in the water and salt balance of the SAR and feeding rivers. Whereas return flows should increase water availability in Iraq, they nevertheless have a negative impact because the concomitant increase in salinity constrains their effective use. This can be clearly noted in the Euphrates, where water salinity has increased significantly at Hit in Iraq due to drainage water from saline irrigation fields (Rahi and Halihan, 2010). To protect the irrigation projects located downstream, drainage water is diverted away from the riverbed of both Euphrates and Tigris rivers through the MOD. Thus, return flows in Iraq are not considered when estimating available water volumes, as they are in fact unusable.

3.6 Water security of the region: conflict versus cooperation

Dating back to 10,000 BC, several civilizations waxed in the Mesopotamia (Iraq), served by the water of both the Tigris and Euphrates rivers. Historically, Iraq was the first country in the Tigris and Euphrates river basins to develop hydraulic structures offering control of extensive flood events and serving irrigation practices. As water represents the source of life, it is also the medium and cause of conflicts. Water disputes in the region date back to 6000 yr and since then, a number of agreements have been reached (Altinbilek, 2004; UN-ESCWA and BGR, 2013). However, all these agreements are bilateral conventions and still there is no agreement involving all parties.

Table 3-3. Water shortages with full development scenario in the Tigris and Euphrates River Basins anticipated after 2040 $(10^9 m^3/y)$.

Basin	Country	Inflow	Tributary	Return flow	Supply	Demands Irrigation	Demands municipal	Demands Evaporation	Balance	Deficit to the SAR
Euphrates	Turkey		28.5	4.7	33.2	-18.0	-0.9	-6.4	7.9	
	Syria	7.9	3.2	2.8	13.9	-7.2	-0.8	-0.6	5.3	
	Iraq	5.3	0.7	1.1	7.1	-30.7	-5.6	-3.9	-33.1	
	Total		**32.4**	**8.6**		**-74.1**				**-33.1**
Tigris	Turkey		20.3	0	20.3	-6.4	-0.4	-0.5	13	
	Iran	13	6.2	0	19.2		-6.2		13	
	Iraq	13	21	3	37	-46.1	-6.3	-4.9	-20.3	
	Total		**47.5**	**3.0**				**-70.8**		**-20.3**

Notes: Future water demands include agricultural uses, municipal uses and evaporation losses, based on the assumption that all planned projects will be operational after 2040 (e.g. Kolars, 1994; Kliot, 1994; Altinbilek, 2004). The total volume of required water for irrigation is calculated based on potential irrigation lands that can be developed. The data for estimating irrigation demands were collected from Kolars and Mitchell (1991), Kolars (1994), Beaumont (1998), and FAO (2009). The estimates on future municipal water needs are based on population projections available from UN-ESCWA-BGR (2013) and The World Bank (2014). Domestic water use rates were obtained from FAO (2007).

Iraq claims a historical right to the water use of both Euphrates and Tigris rivers. The strong disagreement of Iraq concerning the construction of dams in upstream Turkey and Syria reflects its fear of losing water to the upstream riparian countries. Similarly, high tensions between Syria and Iraq emerged in 1975 due to the reductions in flows to Iraq when Syria started filling Lake Assad behind the Tabqa dam on the Euphrates River. The imminent conflict was avoided by the efforts of both countries and regional and international mediation (MacQuarrie, 2004).

Water storage to fill the reservoir created by the Ataturk dam on the Euphrates River in Turkey again raised tensions in 1990 and threatened the water security of the region. Iraq and Syria protested against the cut-off of water by Turkey for one month, which caused considerable harm. Turkey committed to release more than the guaranteed 500 m^3/s based on the 1987 protocol between Turkey and Syria. Turkey also claimed that the reservoir filling time was chosen when the demands were at their lowest, in order to minimize harm to downstream countries (Biedler, 2004; MacQuarrie, 2004).

In the light of the data presented in this study, the potential and looming disagreements and disputes over water allocation between the riparian states may materialize in the absence of a comprehensive basin agreement. Instead of continuing to exploit the natural resources of the basin unilaterally or bilaterally, there is still an opportunity to shift towards an integrated approach. There are a range of possibilities for cooperation between the parties involved to achieve equitable water use and distribution, as well as reducing water pollution.

For example, the energy–water trade may be an important option that can enhance economic cooperation among the riparian countries (Akanda et al., 2007). This may improve communication and cooperation on water management issues. The opportunities created by virtual water trade (Merrett et al., 2003) could inform the revision of the existing water development strategies in the region. Biedler (2004) suggested a river basin management model as the best solution for the region's future water security, based on the principle of utilizing the available waters in a reasonable and equitable manner. An external mediator could initiate the negotiations that could lead to a basin-wide agreement (Biedler, 2004; Akanda et al., 2007).

A comprehensive water resource agreement could enhance regional socio-economic development, deal effectively with increasing water demands, provide a buffer for extreme flood and drought events and improve water quality, sharing the available technologies and other economic incentives between the states. Relatively developed countries such as Turkey could support other riparian countries with improving the efficient and effective use of water. The export of Iraqi oil and gas through Turkey and intensified trade relations can be used to reinforce the cooperation between the riparian countries on water issues. Expanding the industrial sectors to create employment rather than continue the high dependency on the agricultural sector could also reduce water demands as agriculture remains the largest user of water in the region. Among the riparian countries, Iraq needs to take serious measures to improve its water resources management, and develop sophisticated irrigation methods that reduce water use. Groundwater and rain harvesting that have so far hardly been used in Iraq can provide some relief in the near future. Moreover, Iraq needs to explore possibilities of reusing some of the huge amounts of drainage water that are now discharged into the Gulf. The regional development of water- and land-related resources needs to adopt new strategies, based on sharing water and land resource data, including hydrological data, climatic information, data on salinity, return flows, soil types, as well as on water demand and withdrawals. Understanding and monitoring water fluxes between upstream and downstream portions of a river basin are necessary for proper water resource governance (Van der Zaag, 2007). Reliable and timely information is needed for the decision-makers to design effective international treaties and prudent management plans.

3.7 Conclusions

This study concludes that the SAR system is under increasing pressure due to reduction in water quality and inflow quantities coming from the Euphrates, Tigris, Karkheh and Karun Rivers. Additionally, degradation of the Mesopotamian Marshlands also results in reduced flows with high salinity to the river. Regarding water quality, the SAR experiences very high salinity levels, well above the permissible limits of drinking water, irrigation and survival of freshwater fish in most parts of the river. The data shows increasing of several ions which indicate increasing the salinity levels along the SAR. The lowest concentrations of compile ions mainly

Mg, Cl, Na, and Ca during 2014 are in the range of 0.7-2 ppt for the river reach between Qurna and Basra. The highest concentration recorded at the estuary (at Faw) were attributed to the combined effect of tidal influence from the sea, and reduced water inflows and increased salt flows from the rivers upstream. The major causes of decreasing water quantity and quality of the SAR, which are causing many environmental, ecological and socio-economic problems, were identified as: (1) changes in the flow regime and reduced inflows from upstream rivers due to increased rivers' regulations and water uses, (2) decrease in water quality along the major rivers due to polluted return flows from irrigation and wastewater inflows, (3) reduced inflows with increased salinity levels due to high evaporation from the marshes, (4) discharge of high salinity water into the SAR upstream and around Basra city and (5) the effect of increased tidal influence. Mitigation measures should focus on addressing these issues. Long-term integrated water management for the entire southern region must take into consideration available reserves of alternative water sources such as from ground water reservoirs including deep aquifers, desalinization of the waters from the Gulf and saline drainage water, rain harvesting and waste water recycling.

The ongoing problems in the SAR region are likely to worsen over time, as can be inferred from the patterns shown in water resources development and management in the region. The riparian countries, Turkey, Syria, Iraq and Iran, have so far tended to embark on water development planning unilaterally. There are few bilateral agreements between countries facilitating water sharing. The annual water demand of Iraq is higher than the mean water availability. Turkey, Syria and Iran being upstream are in a better position to meet their planned demands (e.g. agriculture, energy and domestic) while Iraq being downstream is likely to face huge water shortages in the future. Thus, this study shows that the problems that the SAR is currently facing, and is likely to continue to face in the near future, cannot be resolved within the SAR region itself. It will require concerted efforts both within Iraq and with the upstream riparian countries.

4 QUANTIFICATION OF SALINITY LEVELS AND VARIABILITY *

4.1 Introduction

Decline in water availability and deterioration of water quality are among the major issues faced by many rivers and dependant human- and ecosystems, most notably in the downstream and delta regions. These changes have serious implications for water security of such regions with visible signs of ecosystem deterioration, and lack of adequate and good quality water for human consumption. These issues are more pressing in semi-arid and arid regions of the world (Vörösmarty and Sahagian, 2000; Canedo-Arguelles *et al.*, 2013).

The SAR, located in the arid environment of Southern Iraq, is such a case where water availability has significantly reduced and the natural regime of the river has drastically changed due to human interventions, i.e. construction of large scale water infrastructure and major water diversions in the upstream basins. River water quality has notably deteriorated due to increasing water withdrawals but also due to untreated wastewater and saline irrigation return flows discharged into the source water bodies. Figure 4-1 shows the soil salinity distribution which varies from slightly saline (1 ppt) at the upstream region to extremely saline (more than 16 ppt) at the delta region. Apart from man-made activities carried out in the upstream part of the basin and along the river, seawater intrusion caused by tidal forces is the main natural driver of salinity variation along the SAR (Rahi and Halihan, 2010; Al-Tawash *et al.*, 2013;

* This chapter is based on Abdullah, A.D., Karim U.F.A., Masih I., Popescu I., van der Zaag P., 2016. Anthropogenic and tidal influences on salinity levels and variability of the Shatt al-Arab River, Basra, Iraq. *International Journal of River Basin Management,* 14(3):357-366 [doi:10.1080/15715124.2016.1193509]

UN-ESCWA and BGR, 2013; Al Mudaffar-Fawzi and Mahdi, 2014; Moyel, 2014; Al-Furaiji *et al.*, 2015; Brandimarte *et al.*, 2015; Shamout and Lahn, 2015).

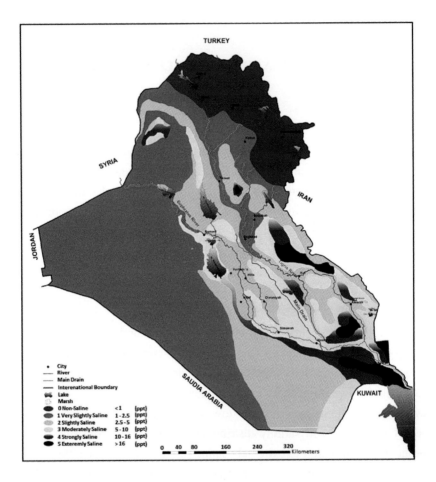

Figure 4-1. Iraq's soil salinity distribution. (*Source: MoWR of Iraq*).

Quantifying in sufficient detail the combined effects of all sources contributing to salinity changes provides the basis for informed water resources management and planning including interventions and investments for improved salinity management. The tidal nature adds to the complexity which warrants a much needed detailed study of the combined effect of the human induced changes along the river and in upstream regions and the tidal influence from the sea (the Gulf) on salinity levels. Therefore, a comprehensive water quality assessment is urgently required for improved water management.

This chapter provides a first set of comprehensive results on the observed salinity dynamics of the SAR based on the analysis of hourly salinity data collected during January to December, 2014 at 10 monitoring stations along the river as described in chapter 2 (see Figure 1-3, Figure 2-2, and Table 2-2 for the locations). The main objective is to quantify the spatial and temporal salinity levels and intra-annual variability. The relative influence of the upstream salinity sources and the seawater intrusion due to tidal phenomenon are also examined.

4.2 Results of salinity analysis

4.2.1 Salinity variations along the SAR

Figure 4-2 presents the summary of the statistical analysis carried out on the observed hourly salinity records for the period 1st January-31st December, 2014. The hourly data of the entire year were 8,730 records at each station grouped into four equal quartiles, the lower quartile starting with the minimum value, and the upper quartile ending with the maximum observed value, and the median value being the value where the two middle quartiles meet. The length between the lowest and highest observed values describes the distribution of the data at each site. The observations revealed a large variability within a year and along the SAR. The salinity was highest at S10 (Faw) with a mean of 17.0 ppt and a range of 0.7-40.7 ppt. The lowest salinity was recorded at S1 (near Qurna town at the Tigris River) with a mean of 1.0 ppt and a range of 0.2-1.6 ppt. Comparison of the central tendency and distribution of the observed data at all stations indicated the possibility of clustering them into distinct spatial units. This allowed the demarcation of the SAR (From Qurna to Faw) into four distinct reaches in terms of the salinity dynamics (see Figure 4-3). This spatial classification could be helpful in understanding the specific salinity behaviour and consequent management interventions along the river.

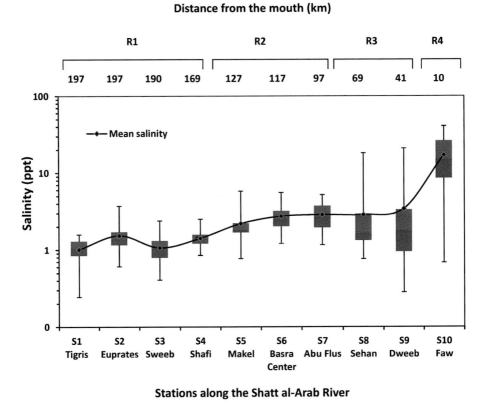

Figure 4-2.*Salinity levels (ppt; log-scale) and intra-annual variability along the SAR at 10 installed monitoring stations during 2014 (distance not to scale).*

The first reach R1 is from Tigris to Shafi and includes the monitoring stations S1 (Tigris), S2 (Euphrates), S3 (Sweeb), and S4 (Shafi). R1 exhibited the lowest salinity levels (mean: 1-1.5 ppt; standard deviation: 0.3-0.6 ppt) compared to the other three reaches. The second reach R2 starts after the confluence of Garmat Ali, from Makel to Abu Flus and includes S5 at Makel, S6 at Basra Centre, and S7 at Abu Flus. The salinity of R2 was notably higher than R1 with mean and standard deviations in the range of 2.2-2.9 ppt and 0.9-1.0 ppt, respectively. The salinity dynamics at reach R3 between S8 (Sehan) and S9 (Dweeb) differed notably from the other reaches. This was evident from the wide range between minimum and maximum values besides the overall higher salinity of R3 compared to R1 and R2. Moreover, the minimum salinity levels at R3 were lower than R2, and similar to R1. The reach R4 represents observations at Faw near the river mouth, where highest salinity levels were recorded (Figure 4-2).

The mean and median values were very similar for all the reaches, except R3. Convergence of the median and mean salinity values in R1 indicate that the salinity was changing smoothly mitigated by the Tigris inflows, whereas in R4 the salinity was increasing and decreasing almost in a uniform pattern due to the effect of the tidal cycle. The larger difference between the mean and the median values in S5, S8 and S9 could be attributed to short-term variations due to human activities such as water releases and water returned from agricultural and industrial practices, coupled with the tidal influence during the year.

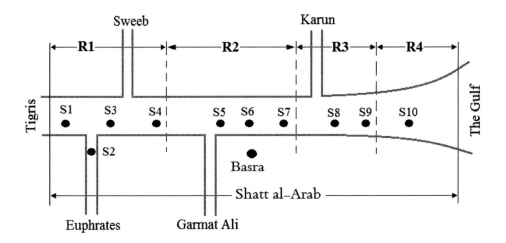

Figure 4-3. *Shatt al-Arab River mainstream schematic plan and its classification into reaches based on salinity dynamics.*

4.2.2 Salinity dynamics at a monthly time scale

The monthly variability appeared to be different at each station and among the monitoring stations. The differences between the reaches were more pronounced than among the stations within a reach. In the case of R1, at the most upstream station (S1 at Tigris) the mean monthly salinity was highest in July (1.4 ppt) and lowest in December (0.3 ppt). The salinity levels fluctuated around 1.0 ppt during most of the months, with the exception of November and December (Figure 4-4a). The salinity levels and variability of S3 at Sweeb were largely similar to that of S1. Station S2 at Euphrates showed a slightly different dynamic with higher salinity levels compared to the other stations, most notable during February to April (Figure 4-4a). On average, the salinity levels were in the range of 1.0 to 1.5 ppt during most of the months, with the exception of February to April when salinity varied between 2.0 to 2.5 ppt. In the case of R4, at station S10 (Faw) located near the river mouth and the Gulf, the salinity levels and variations were the highest compared to all other reaches. The mean monthly salinity fluctuated around 25-30 ppt during June to September, whereas it varied around 10-15 ppt in other months (Figure 4-4c). Notable from Figure 4c, the salinity levels for R3 followed a pattern of variations similar to that of R4, though salinity values at R3 are much lower compared to R4, from 0.5 ppt in December to 11.8 ppt in July along R3. The mean monthly salinity levels in case of R2 varied between 1.6 and 4.5 ppt (Figure 4-4b), whereby two salinity peaks, during February to March and July to September, could be observed.

4.2.3 Daily and hourly salinity changes

The salinity changes within a day are more substantial at R4 and R3 when compared to R2 and R1 (Figure 4-5). Figure 4-5 shows hourly changes in salinities and water levels along the river during two time-periods, one month each, the first during March (spring) and the second during September (summer). In Figure 4-5 one station is presented for each reach, S1, S5, S9, and S10 for R1, R2, R3, and R4 respectively. The semidiurnal tide is the dominating tidal pattern at the river mouth with water amplitude of 1m during neap tide and 3 m during spring tide. The spring

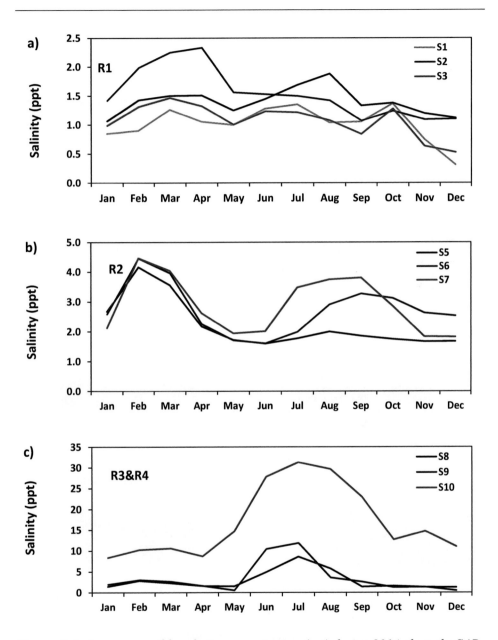

Figure 4-4. *Average monthly salinity concentration (ppt) during 2014 along the SAR (note that y-axes have different scales).*

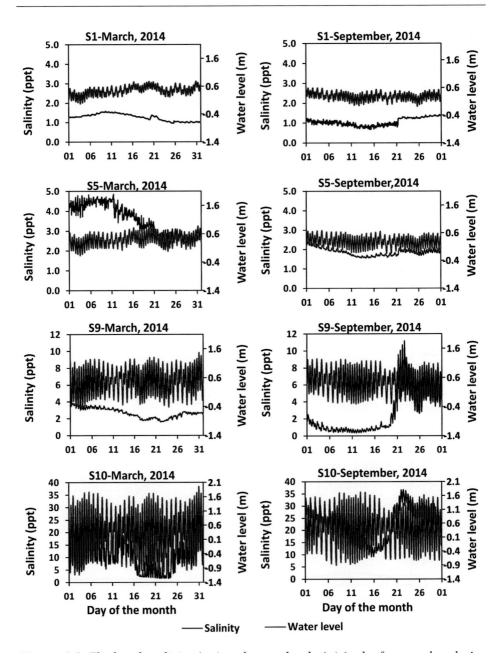

Figure 4-5. *The hourly salinity (ppt) and water levels (m) in the four reaches during March and September 2014 (note that the salinity y-axes have different scales for the different locations).*

and neap tide follow each other in a succession of one week, as can be seen at station S10. The tidal cycle of approximately 12 hours and 25 minutes is observed in the area with notable flood/high and ebb/low tides during this period. The salinity varies notably during spring and summer time at each station and at different locations even though the tidal amplitude remains almost in the same range. Salinity in the upstream portion (R1 and R2) increased during spring time even with relatively high river discharges during March compared to September. Salinity in the downstream portion (R3 and R4) increased during the low flow period as can be seen in September. As expected near the mouth of a river, the salinity at R4 during the flood period was high while it was low during the ebb period as substantiated by Figure 4-5. In general, this correspondence of hourly salinity and water levels is strongest at R4, and gradually becomes weaker for R3 and further upstream. However, at daily time resolution, the salinity levels could be varying differently in contrast to the consistent tidal pattern.

4.2.4 Upstream and downstream relationships

A correlation analysis was conducted to examine the comparative impacts of upstream and downstream salinity sources (Table 4-1). Station S9 at Dweeb exhibited most upstream and downstream influences compared to the rest of the stations (correlations with S8 and S10 were 0.81 and 0.66, respectively). As expected in case of S9, this was due to the tidal influence from the downstream and the dominant Karun River's impact at the upstream. In contrast, S8 showed a more dominant influence from downstream (significant correlation (0.81) with S9), in comparison with the upstream influence (insignificant correlation (0.30) with S7). Two inferences can be drawn from these results. First, the tidal influence was significant at S8 which is located about 70 kilometres from the Gulf. Second, there is a high impact of the Karun River inflow compared to the SAR flow at S7 further upstream. At the other stations, correlations were comparatively weaker and upstream influence was found prominent in these instances. We also conducted multiple linear regressions to predict the salinity levels at an observation point based on upstream and downstream stations, but the results were not promising in most cases. The poor regression results imply that the possibility of reducing the number of monitoring stations is not feasible and more stations may be required to improve the spatial salinity profile along the river.

Table 4-1. *Correlation (r) between neighbouring station records based on hourly salinity observations from 1st January to 31st December, 2014.*

	S1	S2	S3	S4	S5	S6	S7	S8	S9	S10
S1	1.00									
S2	0.25	1.00								
S3	0.71	0.67	1.00							
S4	0.47	0.27	0.45	1.00						
S5	0.11	0.40	0.37	0.08	1.00					
S6	0.03	0.14	0.17	0.01	0.80	1.00				
S7	0.38	0.34	0.48	0.44	0.59	0.67	1.00			
S8	0.38	0.05	0.29	0.57	-0.05	-0.19	0.30	1.00		
S9	0.48	0.04	0.35	0.37	-0.09	-0.28	0.16	0.81	1.00	
S10	0.31	-0.22	0.02	0.44	-0.29	-0.30	0.11	0.65	0.66	1.00

4.2.5 Salinity changes from upstream to downstream

The salinity changes along the SAR were further examined by calculating the change in the mean monthly salinity levels at a station with respect to its upstream station. The results are shown in Table 4-2. It can be inferred from this table that the salinity shows an increasing trend along the river when average annual values are considered. This observation corroborates well with the earlier findings on salinity changes along the Tigris and the Euphrates Rivers where an increase from upstream to downstream is reported (see chapter 3). More specifically in temporal terms, a mixed pattern of increase and decrease was evident (Table 4-2). Associating this pattern to specific seasons was not possible. This is contrary to the expectations that due to the dilution effects the salinity would be less during high flow months (January to April) compared to the normal and low flow months. A dominant increasing pattern could be observed for S4 to S7, with the exception of a few months for S6 and S7. This reflects significant salt addition from local sources within these reaches, mainly irrigation return flows, highly saline water discharged from Hammar marshes through Garmat Ali River and domestic wastes discharged into the SAR, especially at R2. A notable decrease is observed for S8 in most of the months. This was attributed to the inflows from the Karun River. This dilution effect of the Karun gradually

Table 4-2. *The monthly and annual spatial and temporal salinity changes (ppt) at each station compared to the station immediately upstream during the year 2014 (S3 is compared with S1; note that S2 is located at the tributary river).*

	Jan	Feb	Mar	Apr	May	Jun	Jul	Aug	Sep	Oct	Nov	Dec	Total	Annual
S3	0.17	0.46	0.16	0.26	0.01	-0.04	-0.10	0.04	-0.20	-0.06	-0.15	0.70	1.23	0.06
S4	0.08	0.09	0.02	0.14	0.24	0.17	0.39	0.75	0.59	0.08	0.89	1.13	4.57	0.30
S5	1.51	1.92	1.37	0.44	0.38	0.10	0.05	0.06	0.39	0.27	0.39	0.49	7.36	0.58
S6	-0.03	0.07	0.11	0.04	-0.01	0.01	0.12	0.46	0.76	0.79	0.58	0.51	3.41	0.24
S7	-0.18	0.00	0.02	0.16	0.14	0.25	0.75	0.29	0.16	-0.08	-0.30	-0.28	0.94	0.05
S8	-0.32	-0.36	-0.44	-0.39	-0.22	1.44	1.47	0.53	-0.64	-0.44	-0.29	-0.31	0.05	-0.01
S9	0.34	0.07	0.17	0.02	-0.63	1.13	0.38	-0.38	0.87	-0.22	0.02	-0.61	1.15	0.20
S10	3.30	2.38	3.00	4.34	25.25	1.66	1.63	7.26	7.85	9.17	10.14	21.78	97.75	3.90

diminishes downstream as indicated by S8 and S9 due to the salinity contributions from anthropogenic and natural sources. Consistent with the expectation, the highest rate of increase was found at S10 due to substantially higher salinity levels of the seawater which is dominating S10.

4.3 Discussion: factors governing the salinity variations

This section discusses the findings, first by focusing on the four reaches of the SAR, then by looking into the implications for water users, and finally by discussing the seawater intrusion distance.

4.3.1 Salinity changes within reach R1

Salinity changes in R1 are mainly determined by the quantity and quality of water inflows from the Tigris, Euphrates and Sweeb rivers. The review of the available data and studies suggest that the salinity of the Tigris increases along the river course (see chapter 3). At the Turkish-Syrian and Syrian-Iraqi borders, the recorded TDS values were 0.3 ppt on average. The salinity levels increase gradually in Iraq reaching 0.4 ppt at Mosul, 0.8 ppt at Baghdad and 1.0 ppt at Qurna. The main factors governing this rise in salinity levels pertain to the reduction in river flow and discharge of saline waters from the irrigation return flows, groundwater, marshes and wastewater. Given the limited data, it was not possible to quantify the relative contribution of each source. Moreover, monthly estimates were not available to delineate changes in the intra-annual pattern over time, except at station S1.

Similarly to the above findings for the Tigris River, salinity of the Euphrates River is also known to increase along the river course (see chapter 3). For instance, UN-ESCWA and BGR (2013) reported salinity of 0.3 ppt at the Turkish-Syrian border which roughly doubles at the Syrian-Iraqi border. The salinity continues to increase sharply in Iraq and ranges between 2.0 to 3.5 ppt in southern Iraq near the confluence with the Tigris River. The salinity measured at S2 (Figures 4-2 & 4-4) corresponding well with the levels reported for southern Iraq. The highly saline water discharges from irrigation return flows and marshes are the major sources of salinity. The observed data show that the salinity of the Tigris and Euphrates as well the SAR are higher during the irrigation season (November-April), when more return flows are expected. The saline irrigation return flows are considered the biggest polluter of surface water in Iraq (ICARDA, 2012). Cumulative impacts of historical irrigation practices in absence of adequate drainage infrastructures have raised salt concentrations in agricultural soil profiles.

Local irrigation activities along the SAR clearly increased the salinity, through drainage water containing salts washed out from the soil profile. The large agricultural investments in the Mesopotamia plain, the area adjacent to and between the Tigris and Euphrates rivers, contribute to river water salinity and more pressure expected to be added with the planned agricultural expansion.

Inflows from the Sweeb River, which is fed through the Hawizeh marshes, influence the salinity along R1. The salinity of the Hawizeh marshes was in the range of 0.7-1.6 ppt during 2014 (Figure 4-6). These values were close to that of S1 but lower than those recorded at S2. This implies that the Sweeb River does not have a substantial effect on increasing the salinity levels of the SAR along R1. Salinity increases from S3 to S4 ranged from 0.02 ppt in March to 1.13 ppt in December (Table 4-2 and Figure 4-4). These increments could be attributed to the saline waters coming from irrigation return flows, marshes and groundwater. There is a large irrigation scheme serviced by the Shafi canal, which takes in river water mostly during high tide and discharges drainage water during low tide. This is in agreement with Abbass *et al.* (2014) who attributed the salinity to local drainage water. There are no major cities and industries along R1 between S3 and S4, thus, the impact of wastewater is likely to be small.

4.3.2 Salinity changes within reach R2

A considerable increase in salinity was noted in R2 compared to R1. The mean monthly salinity increased by 0.05 in July and 1.9 ppt in February at S5 compared to S4. This increment in salinity levels was due to the saline water flowing from the Hammar Marshes into the SAR via Garmat Ali River. The salinity of the Hammar water ranged between 3.0 and 5.6 ppt during 2014 (Figure 4-6). These values are more than double the salinity recorded at all the stations along R1. Consequently, a sharp rise in salinity levels at the start of R2 was noted compared to the end of R1. The salinity along R2 further increased at S6 (0.07 to 0.79 ppt) and S7 (0.02 to 0.75 ppt) during most of the months (Table 4-2 and Figure 4-4), mainly due to wastewater discharges from Basra city and industries along the SAR. This is in consistent with the findings of Al-Tawash *et al.* (2013) and Al-Manssory *et al.* (2004) who detected industrial and domestic pollution along R2. Contrary to expectations, small decreases in the mean monthly salinity levels (in the range of 0.08 to 0.30 ppt) were recorded for S7 during the period October-January. This decline could be attributed to dilution effect of rain water.

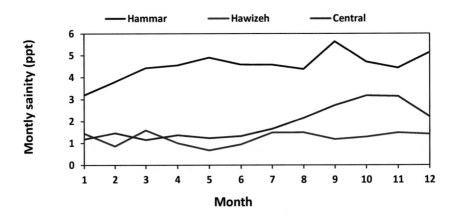

Figure 4-6. *Monthly salinity of marsh water during 2014 (see Figure 1-3 for their locations). Data from the MoWR.*

4.3.3 Salinity changes within reaches R3 and R4

As seen from Table 4-2, there was a notable decrease in salinity at S8 compared to S7 during the period of study except for summer months (June, July and August), when salinity increased in the range of 0.5-1.5 ppt, possibly due to increased tidal impact during this period of low flow (Table 4-2, Figure 4-4). Most of the months showed reductions in the salinity levels in the range of 0.2-0.4 ppt. This decline could be attributed to the impact of Karun, despite limited availability of supporting data. There are few studies describing water quality of the Karun (Afkhami, 2003; Afkhami et al., 2007; Allahyaripour, 2011; Aghdam et al., 2012; UN-ESCWA and BGR, 2013; Adib and Javdan, 2015), but salinity values were not reported or sparsely mentioned in these publications. Nevertheless, Afkhami (2003) reported average salinity levels at Khoramshahr near the confluence with the SAR for the period of 1970-2001. Adib and Javdan (2015) reported salinity levels at Salmaniehe near Khoramshahr for March to May 2001. Recent salinity levels are not available. It can be concluded from Afkhami (2003) that the salinity exhibits large inter-annual variations, roughly in the range of 1.0-4.0 ppt during the reported period of 1970-2001, and are believed to have increased since, due to increased water withdrawals and return flows. River discharge could be a major factor governing this inter-annual variation besides temporally varying contributions from the salinity sources (e.g. irrigation return flows, saline groundwater inflows, wastewater discharge). Although it is not possible to assess the exact salinity levels of the Karun for the period of this study (2014), salinity levels of the Karun water were mostly lower than those of the SAR according to the available information.

The salinity changes at S10 compared to S9 were in the range of 1.7-21.8 ppt. This change was highest when compared to all other stations (Table 4-2). Moreover, salinity levels increased during all the months. The dominance of the tidal phenomenon (seawater intrusion) governed this substantial increase in salinity. Seawater intrusion distance varied along the year which was governed mainly by tidal forces and volumes of river discharges. The observations showed that it can reach up to about 80-100 km (R3 and R2), where the salinity increased at downstream stations 3 to 6 times compare to S6 during June and July (Figure 4-4).

4.3.4 Seawater intrusion length

Here we summaries the above findings by estimating the dominant salinity sources along the SAR: seawater and other sources. The seawater intrusion length (L) is mainly governed by geometrical, hydrological, and hydraulic parameters which determine the mixing and advection processes. Thus a numerical or analytical model is required to predict the maximum L. In this section we introduce a simple method to estimate L preliminarily, by using linear interpolation based on maximum salinity concentrations. For this purpose salinity levels of July were chosen, when the maximum effect of seawater intrusion was recorded. The salinity at the stations along the mainstream are presented according to their distances from the mouth (Figure 4-7). The river reaches can be classified based on their being affected by seawater intrusion. R4 and R3 are those parts directly affected by seawater intrusion, R1 is the unaffected part, and R2 is the mixed part which is affected by the conditions of the two other parts. The plotted values for July (see Figure 4-4) are linearly extrapolated to intersect each other. The salinity of the middle part originates from the mixing of upstream water and downstream seawater with various degrees of influences of the mixture. The intersections of the mixed part with the affected and unaffected parts are found respectively at 80 and 120 km from the mouth. These distances represent the most probable maximum extent of seawater intrusion L at high tide and low river discharge conditions. Further analysis on this subject is done using analytical and modelling approaches and is presented in subsequent chapters (5 and 6).

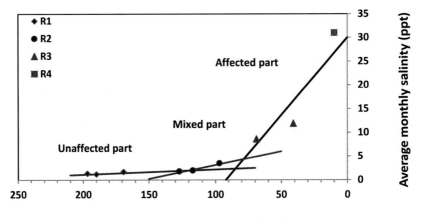

Figure 4-7. *Estimating the seawater intrusion length extrapolated from maximum average monthly salinity records of July.*

4.4 The salinity levels and implications for the users

Water of the SAR is used for agricultural, domestic, and transport purposes along its course. The rising salinity level is a known phenomenon in the area with negative repercussions on agricultural production, livelihoods and the environment. Salinity levels were classified for its suitability for irrigation uses based on the criteria proposed by De Voogt et al. (2000) (Table 4-3). The results on monthly bases are shown in Table 4-4. The SAR water is usable during most of the year along R1. However, a cautious use is warranted near the confluence of the Euphrates with the SAR and further downstream close Shafi station (S4) during part of the year. The current salinity levels of reach R2 belongs 4th and 5th class. There were few months (e.g. June to August) when salinity far exceeded the recommended limits. Water was unfit for irrigation purposes along R3 during part of the year. However, during a few months, probably when high flows were discharged from the Karun, R3 water could be used for irrigation. The salinity levels in R4 were extremely high compared to the standards.

Table 4-3. *Irrigation water quality criteria used in the study according to De Voogt et al. (2000).*

Salinity range (ppt)	Class	Colour coding
0-0.175	1st-Excellent	
0.175-0.525	2nd-Good	
0.525-1.4	3rd-Usable	
1.4-2.1	4th-Use with care	
> 2.1	5th-Unusable	

4.5 Conclusions and recommendations

Monitoring the SAR at an hourly time-step during a full year at 10 carefully selected locations (S1-S10) along 200 km proved that salinity is a high-priority issue requiring measures to reduce the high salt concentrations, taking account of their high variability both in space and time. Four distinct reaches R1-R4 were identified based on similar trends of observed salinity dynamics: reaches R1 Tigris-Shafi (stations S1-

S4), R2 Makel-Abu Flus (S5-S7), R3 Sehan-Dweeb (S8-S9), and R4 at Faw (S10). The lowest and highest measured salinity concentrations were 0.2 and 40 ppt with salinity levels increasing, however not uniformly, towards the mouth. The monthly and finer time resolutions resulted in mixed patterns of increasing and sometimes decreasing salt concentrations. The ranges of monthly mean salinity concentrations for R1-R4 were 1-2, 2-5, 1-12 and 8-31 ppt, respectively. These differences can be attributed to multiple factors, including reduction of discharge and deterioration of water quality in the two upstream tributary rivers (Tigris and Euphrates), irrigation return flows, inflows and discharges of domestic and industrial wastewaters, discharge from the marshes, Karun inflows and seawater intrusion. The relative impact of these factors varies along the SAR course in space and time.

Table 4-4. *Classification of the suitability of the SAR salinity levels for irrigation purposes.*

	R1				R2			R3		R4
	S1	S2	S3	S4	S5	S6	S7	S8	S9	S10
Month	Tigris	Euphrates	Sweeb	Shafi	Makel	Basra Centre	Abu Flus	Sehan	Dweeb	Faw
Jan	0.847	1.419	0.987	1.065				1.449	1.946	
Feb	0.901	1.986	1.313	1.427						
Mar	1.262		1.468	1.503						
Apr	1.054		1.325	1.509				1.604	1.633	
May	0.998	1.561	1.004	1.249	1.719	1.705	1.942	1.522	0.560	
Jun	1.281	1.529	1.234	1.447	1.595	1.604	2.012			
Jul	1.352	1.499	1.216	1.689	1.770	1.988				
Aug	1.038	1.421	1.076	1.881	1.996					
Sep	1.053	1.070	0.840	1.333	1.854			1.389		
Oct	1.366	1.241	1.278	1.378	1.746				1.251	
Nov	0.749	1.091	0.634	1.197	1.665		1.834	1.306	1.327	
Dec	0.309	1.103	0.525	1.119	1.670		1.822	1.253	0.485	

Reach R1 experienced three salinity peaks during 2014, with the highest peak during spring, most likely due to the combined effect of upstream rivers and Marshes discharges mixing with domestic, industrial and irrigation return flows. The main cause of observed abrupt salinity rises along reach R2, typified by two high peaks in

2014 (highest in the winter), is most probably from discharge of highly saline waters from the Hammar marshes via the Garmat Ali River coupled with wastewater from domestic urban and industrial activities, possibly exacerbated by increased petroleum production during 2014. A mixed pattern of high-low salinity was observed in R3 with the highest peak occurring during summer. These changes are attributed to the combined impact of upstream sources of salinity diluted by the Karun River, during most of the time in the year, together with incremental impact of the tail of saline water encroaching from the sea due to low freshwater discharges and resulting increased tidal influence. Annual salinity peaks as measured at S1-S10 were used to estimate the tail of seawater intrusion to be in research R3 between S7-S8 at a distance of about 80 km from the mouth. The highest salinity levels at R4 were caused by the dominant seawater impact with some reduction due to freshwater inflows from upstream. Besides the named factors high evaporation rates in this arid environment have contributed to salinity increases in all four reaches.

Other studies supported by international and local water quality standards corroborate our conclusion that the SAR is unfit as a source for drinking or other human and domestic purposes. While urban users are already provided with treated water, poor (especially) rural communities may be at risk. The SAR water is also found to be beyond recommended irrigation limits along most of its course and during most of the year except along R1. Assuming validity of the results beyond 2014, we caution its use for irrigation from February through April and June through August.

This study recommends for further monitoring of point sources of salinity along the SAR and its main tributaries the Sweeb, Garmat Ali and Karun, including wastewater discharges from the main cities, the main drainage canals from irrigation zones, as well as from the main industries especially from oil production. This requires all the related and competent authorities to work collaboratively (water resources, marshes, environmental, water treatment, and sewage departments), using a comprehensive monitoring program instead of individual and sparse observation efforts. If this is done, and the current stations continue to measure, then the precise sources of the problem can be corroborated. The monitoring program should also include the velocity and flow direction to estimate the water discharges and hence contributing into the estimation of the salt loads caused by different sources.

These observations stress the need for addressing the salinity issues through well-coordinated efforts from all the riparian countries (Turkey, Syria, Iran and Iraq). This will entail actions on improving water quality including focus on salinity reduction and considering more water releases for downstream uses along the SAR. Iraq needs to urgently apply salinity management actions to control discharge of the saline water return flows from the irrigation systems within Iraq. Moreover, wastewater treatment also requires accelerated efforts to avoid salinity increase and avert polluted and saline water entering into the water bodies in Iraq. The application of the models could provide useful information in the decision making process. Application of tools and models like analytical techniques, physically based models, and optimization tools for the SAR would provide further insights into the salinity dynamics governed by anthropogenic and natural factors and inform alternative management strategies.

5 ANALYTICAL APPROACH FOR PREDICTING OF SEAWATER INTRUSION *

5.1 Introduction

Discharge of fresh river water into the ocean is closely related to vertical and longitudinal salinity variations along an estuary (e.g. MacKay and Schumann, 1990; Wong, 1995; Becker *et al.*, 2010; Whitney, 2010; Savenije *et al.*, 2013). River discharge also has a noticeable effect on the tidal range primarily through the friction term (Savenije, 2005). A decrease in river discharge into an estuary could increase the tidal range and the wave celerity, and consequent increase in salinity levels (Cai *et al.*, 2012). Upstream developments of large dams and water storage facilities change the nature of river flow and subsequently alter river hydrology and quality (Helland-Hansen *et al.*, 1995; Vörösmarty and Sahagian, 2000). The SAR which discharges through its estuary at the border between Iran and Iraq into the Gulf is facing serious reductions in freshwater inflows upstream and from its tributaries, as well as significant seawater intrusion downstream. The alteration of river discharge also affects the estuarine ecosystem in terms of sediments, nutrients, dissolved oxygen, and bottom topography (Sklar and Browder, 1998). All these problems are strongly featured in the SAR.

The increases in salinity along the SAR, particularly caused by seawater intrusion, have become a threat to the people and environment alike. Generally, seawater intrusion makes the river water unfit for human consumption and unacceptable for irrigation practices. Saline water in the SAR estuary comes from both natural

* This chapter is based on Abdullah, A. D., Gisen, J. I. A., van der Zaag, P., Savenije, H. H. G., Karim, U. F. A., Masih, I., and Popescu, I. (2016). Predicting the salt water intrusion in the Shatt al Arab estuary using an analytical approach, Hydrology and Earth System Science, 20:4031-4042 [doi:10.5194/hess-20-4031-2016].

(seawater intrusion) and anthropogenic sources. Thus, the pattern of the salinity variation is complex because of the dynamic spatial and temporal interaction between salinity sources. Available studies on the SAR identify the escalating pressure of salinity increment and its consequences for water users as well as the ecosystem (e.g. Al-Tawash *et al.*, 2013; Al Mudaffar-Fawzi and Mahdi, 2014), but detailed information on the extent of seawater intrusion under different conditions is lacking. Hence, there is a need to investigate the impact of seawater intrusion among other sources on the river salinity, and to analyse the dynamics of the saline water-freshwater interface for effective water management.

Different approaches have been used to study the relationship between saline water and freshwater in estuaries. Alber (2002) proposed a conceptual model for managing freshwater discharge into estuaries. Wang *et al.* (2011) used an empirical approach, conducting three hydrological surveys along six locations around the Yellow River mouth to investigate the effect of abrupt changes in the river discharge on the salinity variations. Using a numerical model, Bobba (2002) analysed the mechanism of seawater and freshwater flow in the Godavari Delta and found that freshwater withdrawals contribute to the advance in seawater intrusion. Liu *et al.* (2004) applied a 2-D model to estimate the salinity changes in the Tanshui River, showing that the significant salinity increase is a result of reservoir construction and bathymetric changes. A 3-D model was used by Vaz *et al.* (2009) to study the patterns of saline water in the Espinheiro tidal channel. The result indicates that the model underestimated the salinity distributions for high river inflow. Das *et al.* (2012) used a hydrology-hydrodynamics model to examine salinity variations under different water diversion scenarios in the Barataria estuary, and discovered that the diversions have a strong impact on salinity in the middle section of the estuary and minor impact in the upper section.

Analytical approaches describing salinity distribution in estuaries have been used by Ippen and Harlemen (1961), Prandle (1985) and Savenije (1986). An analytical solution is able to provide important knowledge about the relationship between tide, river flow, and geometry of the tidal channel. The 1-D analytical salt intrusion model proposed by Savenije (1986, 1989, and 1993) is considered (see chapter 2), which uses the more natural exponential geometry and requires a minimal amount of data.

The 1-D analytical seawater intrusion model is based on a number of parameters that can be obtained through field surveys. Variables such as Van der Burgh's coefficient (K) and the dispersion coefficient (D_0) are not directly measurable and therefore they are obtained by calibrating the simulated salinity curve to the datasets from the salt intrusion measurements. For this study four measurement campaigns were conducted, mainly measuring salt concentrations and water levels. The measurements took place during the wet and dry periods at spring and neap tides. These were on 26 March 2014 (neap-wet), 16 May 2014 (spring-dry), 24 September 2014 (spring-dry), and 5 January 2015 (spring-wet). The model was applied under different river conditions to analyse the seasonal variability of salinity distribution. Figure 5-1 shows the measurement locations along the river axis.

The aim of this study is to determine the real extent of seawater intrusion into the SAR estuary. This is done by applying the 1-D analytical seawater intrusion model combined with the revised predictive equations for tidal mixing of Gisen *et al.* (2015b). Then the predictive model was used to examine the consequences of changes in river flow for the salinity distribution.

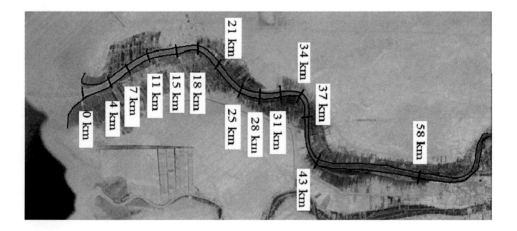

Figure 5-1. *The aerial view of the estuary from Google Earth with the measurement locations (not to scale).*

5.2 Theory of the analytical model

During a tidal cycle, the tidal velocity is near zero just before the tidal current changes direction. This situation is known as high water slack (HWS) just before the direction changes seaward, and low water slack (LWS) just before the direction changes landward. The model originally proposed by Savenije (1989), calibrated with measurements made at HWS, describes the salinity distribution in convergent estuaries as a function of the tide, river flow, and geometry, using Van der Burgh's coefficient (K) and the dispersion coefficient (D_0) at the mouth. A conceptual sketch of the 1-D model of seawater intrusion is shown in Figure 5-2.

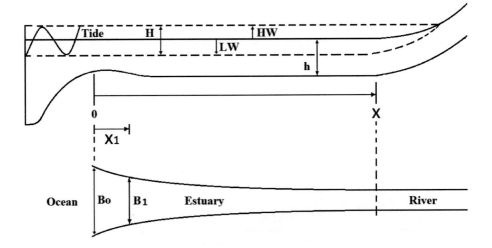

Figure 5-2. *Sketch of the estuary the longitudinal profile and the top view.*

The geometry of an estuary can be presented by exponential functions describing the convergence of the cross-sectional area and width along the estuary as:

$$A = A_o e^{-\frac{x}{a_1}} \qquad \text{for } 0 < x \leq x_1 \qquad (5.1)$$

$$A = A_1 e^{-\frac{(x-x1)}{a2}} \qquad\qquad \text{for } x > x_I \qquad\qquad (5.2)$$

$$B = B_o e^{-\frac{x}{b1}} \qquad\qquad \text{for } 0 < x \leq x_I \qquad\qquad (5.3)$$

$$B = B_1 e^{-\frac{(x-x1)}{b2}} \qquad\qquad \text{for } x > x_I \qquad\qquad (5.4)$$

where A_0 and B_0 are the cross-sectional area [L^2] and width [L] at the estuary mouth ($x=0$), A_1 and B_1 are the cross-sectional area and width at the inflection point ($x=x_I$), and $a_{1,2}$ and $b_{1,2}$ are the cross-sectional and width convergence lengths [L] at $x \leq x_I$ and $x > x_I$, respectively.

Combining (5.1) with (5.3) and (5.2) with (5.4) describes the longitudinal variation of the depth:

$$h = h_o e^{-\frac{x(a1-b1)}{a1b1}} \qquad\qquad \text{for } 0 < x \leq x_I \qquad\qquad (5.5)$$

$$h = h_1 e^{-\frac{(x-x1)(a2-b2)}{a2b2}} \qquad\qquad \text{for } x > x_I \qquad\qquad (5.6)$$

where h, h_0, and h_1 are the cross-sectional average water depths [L] at distance x from the mouth, at the estuary mouth, and at the inflection point respectively.

Integrating the geometry equations into the salt balance equation of Van der Burgh (1972) yields a steady-state longitudinal salinity distribution along the estuary (see Savenije 2005) under HWS condition:

$$S - S_f = S_o - S_f \left(\frac{D}{D_o}\right)^{\frac{1}{K}} \qquad\qquad \text{for } 0 < x \leq x_I \qquad\qquad (5.7)$$

$$S - S_f = S_1 - S_f \left(\frac{D}{D_1}\right)^{\frac{1}{K}} \qquad \text{for } x > x_I \qquad (5.8)$$

where D_0, D, and D_1 [$L^2 T^{-1}$] are the dispersion coefficient at the estuary mouth, at any distance x, and at the inflection point, S_0, S_1, and S [$M L^{-3}$] are the salinity at the estuary mouth, inflection point, and distance x respectively, S_f is the freshwater salinity, and K is the Van der Burgh coefficient which according to Savenije (2005) has a value between 0 and 1, where

$$\frac{D}{D_0} = 1 - \beta_o \left(exp \left(\frac{x}{a_1}\right) - 1 \right) \qquad \text{for } 0 < x \leq x_I \qquad (5.9)$$

and

$$\frac{D}{D_1} = 1 - \beta_1 \left(exp \left(\frac{x - x_1}{a_2}\right) - 1 \right) \qquad \text{for } x > x_I \qquad (5.10)$$

with

$$\beta_o = \frac{K a_1 Q_f}{D_o A_o} \qquad \text{for } 0 < x \leq x_I \qquad (5.11)$$

$$\beta_1 = \frac{K a_2 Q_f}{D_1 A_1} \qquad \text{for } x > x_I \qquad (5.12)$$

B_0 and β_1 are the dispersion reduction rate [-] at the estuary mouth and at the inflection point respectively, and Q_f is the freshwater discharge.

The salt intrusion model is used to estimate the seawater intrusion length, which can be determined using low water slack (LWS, the lower extreme salt intrusion), high water slack (HWS, the upper salt intrusion), or tidal average (TA, the average of the full tidal cycle). Savenije (2012) proposed to calibrate the model on measurements carried out at HWS. This is to obtain the maximum salt intrusion over the tidal cycle. The salinity distribution can be computed at LWS and TA based on the relation between salinity distributions during the three conditions. The salt distribution curve at HWS could be shifted downstream over a horizontal distance equal to the tidal excursion length (E) and half of the tidal excursion length ($E/2$) to obtain the salt distribution curve at LWS and TA conditions respectively. The model variables can be determined from field observations and shape analysis; while the two parameters K and D_0 remain unknown, in addition to Q_f, which is difficult to determine in the tidal region. To facilitate the calibration process, D_0 and Q_f are combined in one variable, the mixing coefficient α_0 [L^{-1}]:

$$\alpha_0 = \frac{D_0}{Q_f} \tag{5.13}$$

After model calibration, the values for K and α_0 are known and the salinity at any point along the estuary can be calculated. Finally the seawater intrusion length (L) during HWS is obtained by:

$$L^{HWS} = x_1 + a_2 ln\left(\frac{1}{\beta_1} + 1\right) \tag{5.14}$$

The calibration parameters can be obtained based on field measurements, but to turn the model into a predictive model, a separate equation for D_0 is required. A predictive equation for D_0 was presented by Savenije (1993) and then improved by Gisen et al. (2015b), who moved the boundary condition to a more identifiable inflection point x_1, based on observations made for a large number of estuaries worldwide as:

$$D_1 = 0.1167 \, E_1 v_1 N_R^{0.57} \tag{5.15}$$

with

$$N_R = \frac{\Delta \rho \, gh \, Q_f T}{\rho \, AEv^2} \tag{5.16}$$

and

$$E = \frac{vT}{\pi} \tag{5.17}$$

N_R is the estuarine Richardson number [-], the ratio of potential energy of the buoyant freshwater to the kinetic energy of the tide, ρ and $\Delta\rho$ [M L^{-3}] are the water density and the density difference over the intrusion length, g is the gravitational acceleration [L T^{-2}], T is the tidal period [T], v is the velocity amplitude [L T^{-1}], and E is the tidal excursion [L].

This study tests the predictive performance of the 1-D analytical salt intrusion model, combined with new revised predictive equations to analyse the real extent of seawater intrusion in the SAR estuary under different river discharge conditions.

5.3 Salinity modelling

5.3.1 Geometric characteristics

Results of the cross-sectional area, width, and depth are presented in a semi-logarithmic scale plot in Figure 5-3. This Figure shows a good agreement between the computed cross-sectional areas A, width B, and depth h based on equations (5.1)-(5.6) and the observed data, except for the part between 40 and 50 km, which is shallower in comparison to the rest of the estuary. The cross-sectional area A and width B are divided into two reaches with the convergence lengths a_1 and a_2 of 22 km and 26 km respectively (see Table 5-1). The geometry changes in decreasing pattern landwards following an exponential function. In an alluvial estuary, the wide mouth and shorter convergence length in the seaward part is generally wave-dominated, while the landward part with longer convergence length is tide-dominated. The average depth h is almost constant with a very slight decrease along the estuary axis (a depth convergence length of 525 km).

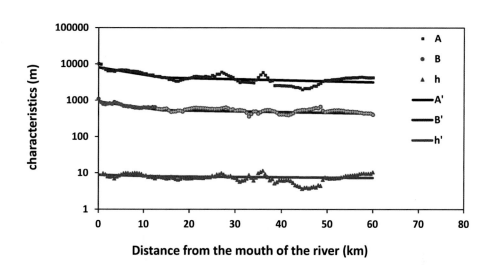

Distance from the mouth of the river (km)

Figure 5-3. *SAR geometric characteristics (A, B, h: measured; A', B', h': equations 5.1-5.6).*

Table 5-1. *The geometry characteristic of the estuary.*

A_0 (m²)	A_1 (m²)	B_0 (m)	B_1 (m)	a_1 (m)	a_2 (m)	b_1 (m)	b_2 (m)	$\bar{h}$ (m)
8,050	4,260	910	531	22,000	160,000	26,000	230,000	7.9

Notes: A_0 and A_1 are cross-sectional areas at the mouth and inflection point respectively. B_0 and B_1 are channel widths at the mouth and inflection point respectively, and a_1, a_2, and b_1, b_2 are locations of the convergence length of the cross-sectional area and width respectively. $\bar{h}$ is the average depth over the estuary length (of 60 km).

5.3.2 Vertical salinity profile

In Figure 5-4 the results of the observed vertical salinity profile at HWS are presented. It can be seen that the salt intrusion mechanism is well mixed for the entire observation period. During the wet period when river discharge is relatively high, partially mixed condition can be observed particularly in the downstream area (Figure 5-4 a and d). In the neap-wet condition as shown in Figure 5a, there is more stratification and the partially mixed pattern occurs in almost the entire stretch of the estuary. This is because at neap tide, the tidal flows are small compared to the high freshwater discharge during the wet season. Conversely, during the spring-dry period when the river discharge is significantly low and the tidal range is large (Figure 5-4 b and 5-4c), the vertical salinity distribution along the estuary is well mixed.

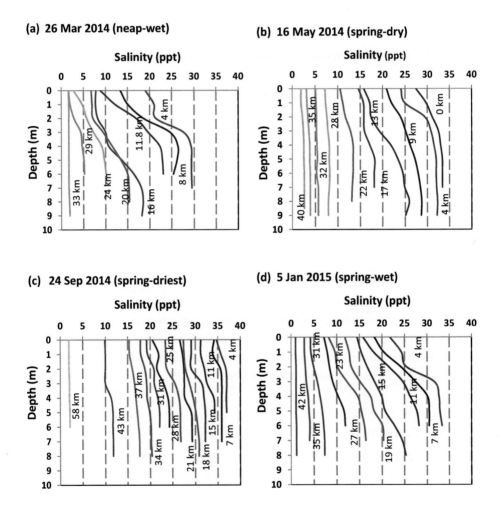

Figure 5-4. *Vertical salinity distribution of the estuary measured between 0 and 58 km at HWS.*

5.3.3 Longitudinal salinity profile

The measurements of salinity during HWS and LWS are presented in Figure 5-5. Calculations of the longitudinal salinity profiles are based on equations (5.7)-(5.14), where the dispersion D decreases over x until it reaches zero at the end of the seawater intrusion length. Coefficients K, D_0, and E were calibrated to obtain the best fit between measured salinity data and simulated salinity variations. The longitudinal salinity distributions during a tidal cycle are demonstrated by three curves: (1) the maximum salinity curve at HWS; (2) the minimum salinity curve at LWS; and (3) the average of HWS and LWS representing the average salinity curve at TA. Tidal excursion (E) is determined from the horizontal distance between the salinity curves of HWS and LWS. This distance is considered constant along the estuary axis during the tidal cycle. In this study, the tidal excursion is found to be 14 km on 24 September and 10 km for the other observations (Table 5-2).

The results show good agreement between measured and simulated salinity profiles with few deviations between the observed and modelled salinities. The small deviations may be due to the timing errors in which the boat movement speed did not coincide exactly with the tidal wave. In Figure 5-5 (a and d), it can be seen that the measured salinity at distances 20 and 24 km during HWS are higher than the simulated values. There is a sub-district (with considerable agricultural communities) and a commercial harbour, and it is believed that all of their effluents and drainage water are discharged into the river. This could be the reason for the salinity being a little higher than expected. In Figure 5-5c, the last measurement point is lower than the simulated one. This may be due to the relatively shallow stretch between 40 and 50 km, which can substantially reduce the seawater intrusion. Also a timing error may be an explanation for this deviation: the boat did not move fast enough as it was delayed for short stops at police checkpoints.

All the field surveys indicate that the maximum salinity at the mouth ranged from 24 to 35 ppt (Table 5-2). The lowest maximum salinity is during the neap-wet period and the highest is during the spring-dry period. It can be seen that the seawater intrudes furthest in September (spring-driest period) and shortest in March (neap-wet). These findings are logical because during the wet season, the estuary is in a discharge-dominated condition and the lower tide (neap) can be easily pushed back by the river discharge. On the other hand, during the dry season the estuary is tide-dominated and the higher tide (spring) managed to travel further inland without much obstruction (low freshwater discharge). The tidal ranges recorded during field surveys are 1.7, 3.2, 2.1, and 2.6 respectively as same date shown in Figure 5-5 (a)-(d).

Besides seawater intrusion, human activities in the upstream part of the estuary also contribute to the salinity levels along the river. From observations, the river salinity in the inland part varies in space and time between 1-2 ppt. Thus, the salt

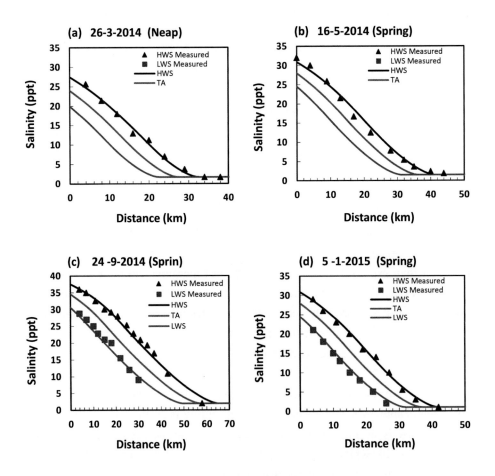

Figure 5-5. *Predicted and measured salinity distribution during HWS, TA, and LWS.*

concentrations are the result of a combination of anthropogenic and marine sources. The findings from the longitudinal salinity distribution indicate that there is a need to analyse and classify the effects of natural and anthropogenic factors on estuary salinity.

Table 5-2. *Characteristic values of the estuary including the maximum salinity at the mouth S_o, the river discharge Q_f, tidal excursion E, Van der Burgh coefficient K, the dispersion coefficient D_o, mixing number α_o, and seawater intrusion length L.*

Period	S_o	Q_f	E	K	D_o	α_o	L
	(ppt)	(m³/s)	(km)		(m²/s)	(m⁻¹)	(km)
26 March 2014	24	109	10	0.65	403	3.7	32
16 May 2014	28	91	10	0.65	473	5.2	42
24 September 2014	34.6	48	15.5	0.65	442	9.2	65
05 January 2015	28	53	10	0.65	281	5.3	42

5.3.4 The predictive model

The dispersion coefficient D is not a physical parameter that can be measured directly. It represents the mixing of saline water and freshwater, and can be defined as the spreading of a solute along an estuary induced by density gradient and tidal movement. Knowing the river discharge is crucial for determining a dispersion coefficient D from Eq.(5.9). However, it is difficult to measure the river discharge accurately in the tidal region due to the tidal fluctuation. In this study, the river discharge data on the days of the measurements were used from the gauging station located at the most downstream part of the river network.

For the situation where measured salinity is known, the dispersion coefficient D_0 and seawater intrusion length L at HWS were calibrated by fitting the simulated salinity curve (Eqs. 5-7 to 5-14) against the field data. In case no field data are available, the dispersion coefficient D_1 was estimated using Eq. (5-15). The predicted D_1 then was used to determine the predicted D_0 (using Eq. 5-9) and L (using Eq. 5-14). Comparisons between the calibrated and predicted values were done to evaluate the performance of the model. The prediction performance was evaluated with two model accuracy statistic: the root mean squared error (*RMSE*) and Nash-Sutcliffe efficiency (*NSE*). The index (*NSE*) ranges from -∞ to1. It describes the degree of accurate prediction. An efficiency of one indicates complete agreement between predicted and observed variables, whereas an efficiency of less than zero indicates that the prediction variance is larger than the data variance.

Figure 5-6 presents poor correlations between the calibrated and predicted values of *D;* the situation is better in the case of *L* values. Table 5-4 displays the correlation between predicted and measured values. The *NSE* obtained for *D* is -0.09 and reflects weak predictive performance. Generally the model appears to overestimate the values of the dispersion coefficient compared to the calibrated ones during the wet period and to underestimate the value during the drought period in September. This could be due to the use of the measured discharge at the end of the tidal domain, which gives higher or lower values than the exact freshwater discharging into the estuary, as it does not account for the discharge of the Karun River at the downstream end and the water consumption and water losses within the system (see Table 5-3). The SAR is the main freshwater source for irrigation, domestic and industrial activities in the region. Hence, water consumption could highly affect the performance of a predictive model especially in the region where water withdrawals can considerably reduce river discharge into the estuary.

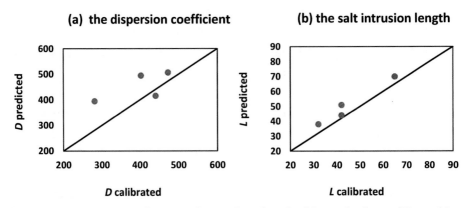

Figure 5-6. *Comparison between the predicted and calibrated values of D_0 and L.*

In order to reduce the uncertainty in the discharge data, some alternative approach has to be adopted. Gisen *et al.* (2015a) estimated the discharges for the downstream areas by extrapolating the correlation of the gauged area with the ungauged areas. Cai *et al.* (2014) developed an analytical approach to predict the river discharge into an estuary based on tidal water level observations. This method is only applicable in estuaries with a considerable river discharge compared to the tidal flows. In this study, a simple approach has been used to assess the discharge in the SAR estuary by deducting the water withdrawals in the downstream region from the discharge data collected at the lowest gauging points. In a similar way the average discharge of the Karun was also estimated (Table 5-3). Data on water withdrawals were collected from the water resources authority and water distribution departments. Besides irrigation and domestic supply, the industrial sector, including the oil industry, is also a significant water user. Unfortunately this study could not obtain information on water usage and disposal by the oil industry.

The adjusted river discharge data are then applied in the predictive model to evaluate the improvement of these changes in predicting values of D and L. The results obtained after the adjustment are shown in Figure 5-7. The figures demonstrate the improvements in predicting the dispersion and maximum seawater intrusion length and show the importance of computing the freshwater discharge accurately. Furthermore the correlations between predictive and observed values are improved for both D_0 and L, 0.46 and 0.9 respectively, and the RMSE also reduced to 60 m^2/s and 4 km for D_0 and L, respectively (Table 5-4).

Table 5-3. *Measured and adjusted river discharge considering water consumption on the days of measurements.*

Date	Measured river discharge (not counting water abstractions and excluding the Karun inflows) (m^3/s)	Adjusted river discharge (deducting water abstractions and including the Karun inflows) (m^3/s)
26 March 2014	109	114
16 May 2014	91	96
24 September 2014	48	58
05 January 2015	53	63

Table 5-4. *Results of the model performance in terms of root mean squared error (RMSE) and Nash-Sutcliffe efficiency (NSE).*

		NSE	RMSE
Measured river discharge	D_0	-0.09	76 m^2/s
	L	0.75	6 km
Adjusted river discharge	D_0	0.46	60 m^2/s
	L	0.9	4 km

The prediction performance of the model is demonstrated in Figure 5-8, where the salinity curves were computed from the predictive equation of D_1 and the adjusted river discharges. Figure 5-8 shows that the prediction salinity curves perform very well compared to the calibrated one during all periods, except January 2015. This could be because the use of the average discharge of the Karun which has a lower than the actual discharge during January, is in the middle of the wet season. At such a time the SAR is expected to receive high flow rates from the Karun River. On the other hand, during this season more return flows are drained into the SAR from the large irrigation scheme serviced by the Karun water system, increasing anthropogenic salinity levels. Accurate estimation of river discharge into the estuary is important in improving the predictive skill of the model.

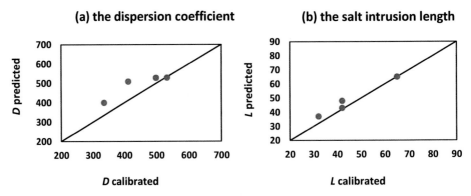

Figure 5-7. *Comparison between the predicted and calibrated values of D_0 and L using the improved discharge data.*

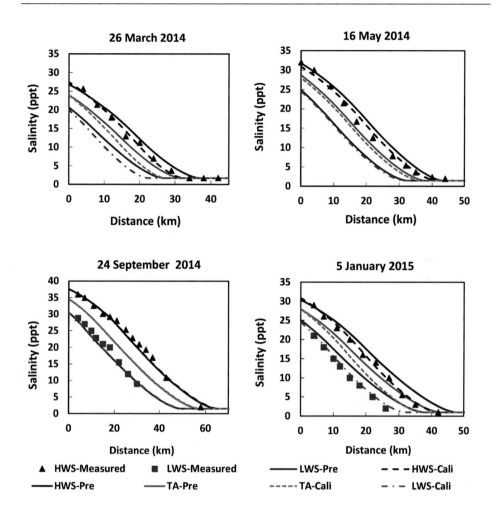

Figure 5-8. *Comparing the salinity curves of the calibrated results (dashed lines) and the predicted results (solid lines) to the observed salinity during the four periods of 2014.*

The ultimate objective of the modelling is to assess the influence of upstream development on the estuarine environment, and also to find the real extent of seawater intrusion. The salinity distribution along the estuary is highly linked to upstream conditions, such as flow regulation and water withdrawals. For the purpose of improving the SAR estuary management, the approach model can lead to estimation of a seawater intrusion length for a given freshwater discharge. This is useful for water supply managers to determine the appropriate location (salinity-free region) for water intake stations. In Figure 5-9 demonstrates the seawater intrusion length (L) associated with different river discharges is plotted corresponds to water released from the Tigris River into the SAR. The seawater intrusion lengths are plotted against a range of freshwater discharge from 5 to 120 m³/s. The main finding is that the length of seawater intrusion increases in a non-linear way with decreasing river discharge. The seawater intrusion length is very sensitive to river discharge when the flow is low. From the plot it can also be seen that the maximum seawater intrusion could reach 92 km from the SAR estuary at 5 m³/s river discharge. This outcome exceeds a preliminary estimated (presented in chapter 4) based on a one-year data series, where the seawater intrusion was estimated to reach up to 80 km considering the annual salinity peaks along the river. An 80 km intrusion length corresponds to a measured river discharge of 58 m³/s, whereas for the predictive model this distance corresponds to a much lower discharge (7 m³/s). It is, however likely that the true river discharge was lower, since during the lowest discharge the irrigation demand is relatively largest. It should also be realized that in the region of 40-50 km the depth and cross-sectional areas are substantially less. Such a shallow reach can reduce the seawater intrusion length substantially, as can be seen from Eq. (5-14), where β_1 is inversely proportional to A.

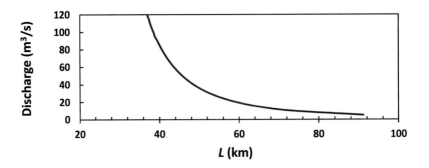

Figure 5-9. *Relationship between river discharge and predicted* seawater *intrusion length.*

5.4 Conclusions

A 1-D analytical salt intrusion model was applied for the first time to the SAR estuary based on four survey campaigns in 2014 and 2015. This model is used to determine longitudinal salinity distribution and the length of seawater intrusion. The analytical model is shown to describe well the exponential shape of the estuary in the upstream direction. Moreover, the results show good agreement between computed and observed salinity under different river conditions. This indicates that the analytical model is capable of describing the extent of seawater intrusion along the SAR estuary.

Results for the dispersion coefficient D_o indicate that the measured river discharge out of the tidal range is higher than the real discharge into the estuary. This can be attributed to water withdrawals along the tidal domain. In the case of low river discharge, water withdrawals have a considerable effect on the predicted seawater intrusion length. The river discharge into the estuary was revised considering water withdrawals for the irrigation and domestic sectors. Use of adjusted river discharge improved the performance of the predictive equations. For further improvement, it is recommended to obtain more accurate estimation of the river discharge into the estuary.

Seawater intrusion is driven by the discharge kinetics from tidal seawater and the hydrostatic potential energy from freshwater fluctuations. Intrusion lengths of 38, 40, 65, and 43 km correspond to tidal ranges of 1.7, 3.2, 2.1, and 2.6 m during March 2014, May 2014, and September 2014, and January 2015, respectively. The longer seawater intrusion distance is caused by low river discharge, as evident for September (dry period).

The SAR is the main source of freshwater for daily consumption and irrigation. Decreased freshwater discharge and increased seawater intrusion will exacerbate an already critical situation in that important agricultural and ecological region. The model shows a scenario in which decreasing river discharge, considered a likely event, can result in an increase in seawater intrusion further upstream to a distance of 92 km. Additional salinity sources from anthropogenic activities will diminish the volume of freshwater leading to very serious health problems and water and food insecurity. Calibration of the model can be enhanced with further monitoring of discharge and salinity from all the tributaries and used to make new estimates of longitudinal salinity distribution under extreme conditions. Preventing seawater intrusion of these magnitudes can only be achieved if the water quantity and quality of the upstream sources as well as along the SAR are promptly and strictly regulated.

6 IMPACTS OF COMBINED SALINITY SOURCES ON THE WATER RESOURCES MANAGEMENT *

6.1 Introduction

Water quality degradation is the main cause of reduced water availability and consequently reduces its use. Increasing salinity is a major water quality problem in many rivers in the world; in particular in downstream and delta regions, often associated with intensive human activities (Thomas and Jakeman, 1985; Shiati, 1991; Roos and Pieterse, 1995; Peters and Meybeck, 2000). According to WHO (1996) at salinity levels greater than 1 ppt water will become undrinkable; and at levels above 3 ppt water becomes no longer suitable for most agricultural uses. Irrigation with high saline water causes yield reduction (FAO, 1985; Rahi and Halihan, 2010). Water salinity is also a major factor affecting estuarine ecosystems. The salinity distribution reflects the biota habitat condition of an estuary (Jassby *et al.,* 1995).

Considering a combination of different salinity sources is of crucial importance for improved salinity managements rather than examine of individual factors. Therefore, there is a need to study the factors that determine the salinity of rivers under tidal influence, including irrigation practices, industrial effluents, urban discharges, quality and quantity of upstream river inflow, and seawater intrusion. This will provide a scientific basis to explore the effective measures for controlling water salinity and resource management to ensure sustainable development.

The SAR is considered the main surface water resource for Basra, the second major city in Iraq. The rise in salinity of the river water is due to natural and

---

* This chapter is based on Abdullah A., Popescu I., Dastgheib A., van der Zaag P., Masih I., Karim U., submitted. Analysis of Possible Actions to Manage the longitudinal Changes of Water Salinity in a Tidal River. Submitted to *Water Resources Management.*

anthropogenic sources and has increased the salt content in the soil and deforested the date palms along its banks. The SAR is experiencing environmental crisis as a consequence (Hameed and Aljorany, 2011; Maser et al., 2011; Abdullah et al., 2015). As from 1980, little effort has been made by researchers and policy makers to comprehensively study and address the issues related to river salinity.

Recently, some environmental studies were conducted in the Shatt al- Arab region, mainly of biological (Al-Meshleb, 2012; Ajeel and Abbas, 2012) and chemical nature (Al-Saad et al., 2011; Mohammed, 2011). Limited investigations were made of the physical processes in this region influencing the dynamics of salinity. Understanding salinity changes driven by natural and anthropogenic sources is essential to provide the basic step for predicting salinity dynamics and better water allocation strategies. Longitudinal salinity distribution is highly dependent on a combination of salinity sources that are controlled by freshwater and brackish water discharges and tidal forces. Releases from the marshes and a number of tributaries into the system with different discharge and salinity concentrations complicate further the problem. A high spatial and temporal gradient requires numerical model support to simulate such a complex water system.

The study presented here uses the 1-D version of Delft3D. This hydrodynamic simplification is used to investigate only the longitudinal dimension along which water salinity changes and salinity sources are simulated. This study is to understand the changes of salinity regime that are caused by combined factor impacts. The study looks at the effect on salinity gradients by combining impacts of tides, return flows and domestic effluents, while having to deal with scarcity of data associated with discharges and salinity levels of different sources. The model has been developed to correlate water flows and salinity concentrations along the SAR taking into account the tidal effect of the Gulf at the mouth of the river. The main purpose of the study is to investigate the salinity dynamic along the river due to different water resources management strategies, by:

- evaluating daily and seasonal effect of a combination of salinity sources on the longitudinal salinity changes which determine the salinity gradient along the estuary;
- comparing the effect of both river flow and tidal forces on salinity changes;

- evaluating the capability of the numerical model to predict salinity changes driven by natural and human activities;
- analysing some water management scenarios;
- exploring and determining the most influential parameters among the tide and river characteristics influencing salinity changes.

This chapter presents the results of the numerical simulations, and describes the impact of different management actions on salinity gradients as they were investigated using the selected model.

6.2 Material and method

6.2.1 Available data

Field data used for this study are hourly time series of the water level, temperature and salinity for the entire hydrological year of 2014, as described in detail in chapter 2. The survey campaign was conducted to provide information and data on water levels and salinity indicators along the SAR taking into consideration the tidal effects. The survey consisted of 10 monitoring stations (Figure 1-3). These stations were installed upstream and downstream of each identified water salinity source. Currently the Tigris River is the main source of the freshwater feeding the upper course of the river. The SAR unites with the Karun River, the main tributary in its lower course, near Abadan city before discharging into the Gulf which is completely located in Iran. Though the water inflow and salt transport from the Karun are important elements characterizing the salinity pattern of the SAR, no data were available on the Karun River for this study. Daily time series of the Tigris River discharge were obtained from the water resources authority in Basra. River cross sectional data were based on a survey carried out in 2012 by the GDSD (General Directorate of Study and Design).

The irrigation system of the SAR is highly linked to the tidal frequency of the Gulf, especially in the lower course. Several canals are used to provide irrigation water to the surrounding farms at high water level of the SAR tidal cycle, returning draining flows at low tide. This makes it difficult to estimate water consumption and return flows accurately. There is no accurate information on water withdrawals and return flows from the Iranian side for the Karun River and some irrigation canals.

The estuary experiences a tidal cycle of approximately 12 hours 25 minutes with notable flood and ebb tides (Figure 6-1). The estuary has a mixed-diurnal and semi-diurnal tide with successive spring and neap tides. The tidal range (the difference between the water levels at high water (HW) and low water (LW)) varies from 1 m (neap) to 3 m (spring). The observed tidal propagation for the last four days of January 2014 at the most downstream (S10) and most upstream (S1) stations is presented in Figure 6-1. There is a large decrease in amplitude from S10 to S1 with a phase lag of around 8 hours. The average water level at the downstream end is around 0.4 m ranging from -1.33 to 1.96 m, while at the upstream it is around 0.34 ranging from -0.34 to 0.84 m. The low waters take a longer time than high waters (the duration of ebbing is longer than flooding), especially in the downstream area. Salinity levels fluctuate at an hourly scale depending on the tide cycles and freshwater discharge. Salinity increases during flood tides and decreases during ebb. The impact of freshwater inflows can be clearly recognized during neap tide and ebb periods. The salinity level also varies along the year, for example the highest value measured in the estuary in 2014 was 40 ppt during summer and the lowest value was 0.7 ppt.

6.2.2 Model set-up

A Delft3D hydrodynamic model was developed and used in this research (see chapter 2). The present study is based on the analysis of the collected data and available secondary data described in chapter 3 and 4. Using 2-D or 3-D numerical models to simulate salinity changes requires sufficiently fine grids implying intensive computational time compared to a 1-D model. Furthermore, the SAR is considered a well-mixed estuary and the transport mechanisms are dominant in the longitudinal direction. Therefore, 1-D model setup was chosen. This is sufficient to simulate salinity changes along the length of the river course including along its estuary. Daily and seasonal salinity variations and solution time step of 10 minutes are used for testing different management scenarios. This 1-D representation purposely serves to identify the bigger picture from which optimization studies aid an overall best-practice decision for the large-scale problem.

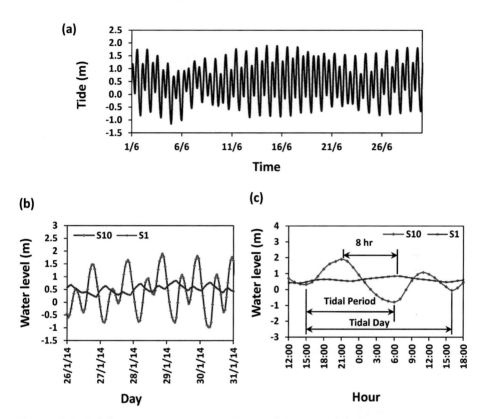

Figure 6-1. *Tidal properties at station S10 and S1: a) Tidal elevation at station S10 in June 2014, b) tidal elevation during 26-31 January 2014, c) phase difference during 28-29 January 2014.*

6.2.3 River geometry

The bathymetry of the riverbed has been determined using the average depth at every cross section at 500 m intervals over the longitudinal axis. The cross-sectional area over the average water depth determines the width of the grid. The number of discretization cells on the longitudinal axis was 388, and the cell size was 300×500 m. The river was extended in the upstream direction to a non-tidal area resulting in a computational domain length of 221 km (Figure 6-2).

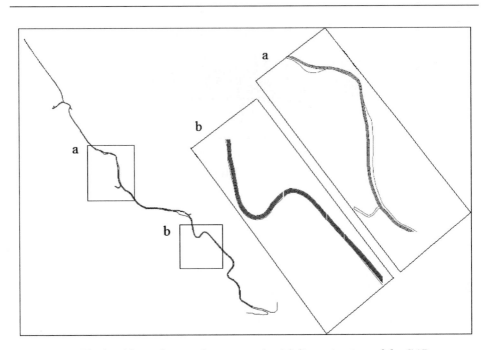

Figure 6-2. *The land boundary and structured grid discretisation of the SAR.*

6.2.4 Boundary conditions

The Tigris contributes the majority of freshwater volume (30 to 100 m³/s) discharged into the SAR at the upstream portion. Therefore, the Tigris is treated as the mainstream of the system, while the other rivers are treated as tributaries. The water contributions into the SAR from each of the other regulated tributaries (except the Karun) ranged between 0 to 10 m³/s (estimated by the local water resource authority). This is a relatively small value compared to the tidal flux and Tigris discharge. The Tigris River discharge was measured at the most downstream water regulator, located out of the tidal range. Time series of daily river discharges for the year 2014 are specified at the upstream boundary, and hourly measured tide elevations near the mouth for the year 2014 are specified as the downstream boundary. The water levels and salinity concentrations during the first time step of the simulation periods at all station were specified as the initial conditions.

6.2.5 Model calibration

Water level variation is substantial at the mouth of the river and it is influenced by tidal amplitude, however this influence gradually becomes weaker along the estuary and further upstream (see Figure 6-1). Moreover, at the lower portion of the river salinity is driven by tidal impact, while at the upper portion it is driven by river discharge (as described in chapter 4). Tidal amplitudes and salinity changes highly are correlated at the downstream portion. Therefore, water elevations are calibrated for the entire 2014 considering only two stations (S8 and S9) as these are located in the lower portion of the river, where the tide is more dominant.

Several irrigation canals mainly in the downstream part branched from the SAR, amounting to 140 canals on each bank. During the flood period of a tidal cycle, water is extracted from the river, while during ebb water returns back by these canals into the river. To simulate this process, the water level is calibrated using two operation points considering water consumption and water losses of the system. A comparison of modelled and observed water levels at S8 and S9 is shown in Figure 6-3. Two periods have been chosen to present the model results, consisting of 8 days each, including spring and neap tide during February and October respectively. The results indicate a good agreement between the simulated amplitudes and phases of tidal elevations with observations. To assess the accuracy of the model outcomes, a quantitative evaluation was used to compare the predicted and measured values. The method of root mean squared error (RMSE) was used to define the accuracy of the model. Another statistical index of agreement, the coefficient of determination (R^2), was used to measure the correlation between the simulated and observed water levels.

R^2 values of 0.82 and 0.85 at S8 and S9 respectively show that the simulated water levels correlate well with the observed water levels. The *RMSE* of the tidal levels are 0.16 and 0.19 m at S8 and S9 respectively. These could be attributed to two main reasons; firstly, the measured water elevations may have been affected by the influences of wind and navigation along the river, and secondly, the operation processes of the river system could affect the tidal amplitudes and phases.

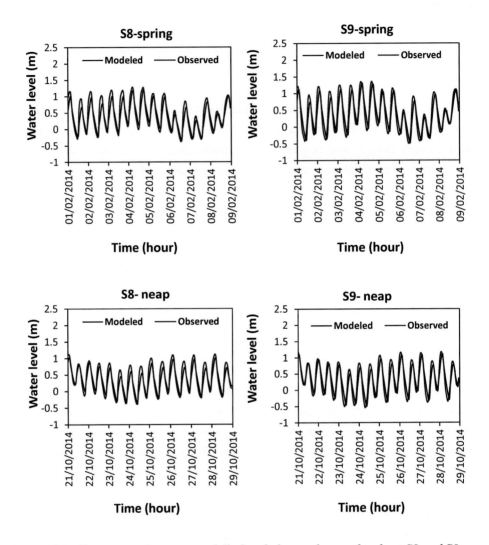

Figure 6-3. *Comparison between modelled and observed water levels at S8 and S9.*

The calibrated hydrodynamic model is expanded with salinity. This is done by adding the hourly series of measured salinity at S1 to the river discharge at the upstream, and measured salinity at S10 to the tidal elevation at the downstream boundary. Salinity from other sources including return flows, domestic and industrial effluents along the river is added to the calibrated water levels as open boundaries at the corresponding locations. Salinity caused by the Euphrates and Garmat Ali rivers and connected marshes is represented by the measured salinity at S2 and S5 respectively. Station S2 recorded the salinity of the Euphrates at 3 km from its confluence point with the SAR. Therefore, the model was first developed for the upper part only to determine the salinity of the Euphrates at the confluence point. The salinity caused by the Shafi and Abu Flus irrigation schemes and surrounding communities are represented by measured salinity at S4 and S7, while the salinity at S8 reflects the impact of the Karun River. The model was verified with S3, S6, and S9 data for the entire year 2014. The water salinity shows large variability within a day and between days. Figure 6-4 shows a comparison between measured and simulated salinity for S3, S6, and S9 at the upstream, middle, and lower portion of the river respectively. Generally, the results reflect a good agreement between the simulated and measured salinity for all those stations despite a few deviations that can be explained.

The model could not capture some of the low salinity concentration as can be seen in S3 during a few days in September and in S9 during a few days in April. Also the model under estimated the salinity at S6 during a few days in February. This could be caused by local fresh or saline water inflows. S3 and S9 recorded salinity levels lower than its upstream and downstream stations during certain periods; this could be attributed to local freshwater discharges from the Sweeb River in the case of S3 and the Karun system in the case of S9. The high salinity concentration at S6 during February could be caused by wastewater effluents, whose impact increased during rainstorm events. Generally, the salinity distribution is highly affected by the operation processes including water withdrawals for domestic, industrial, and irrigation use, and the quantity and quality of return flows. The *RMSE* of salinity concentrations to the observed ones are 0.2, 0.5, and 1.4 ppt, and the calculated R^2 are 0.78, 0.84, and 0.89 for S3, S6, and S9 respectively. The good correlation between the simulated and observed salinity implies that the model can reasonably simulate salinity changes caused by combined terrestrial and marine sources.

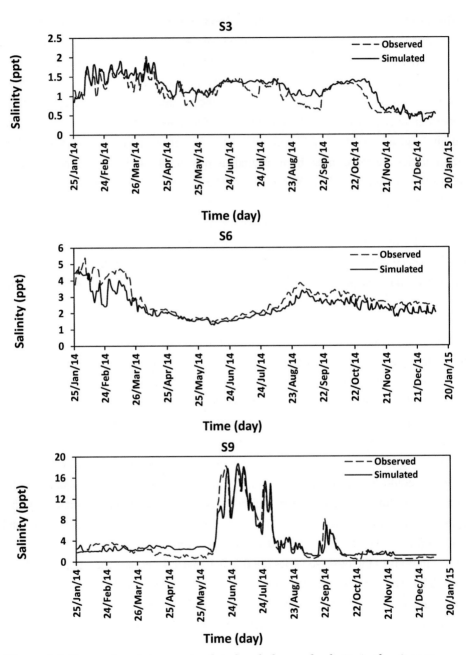

Figure 6-4. *Comparison between simulated and observed salinity in the river at stations S3, S6, and S9.*

6.2.6 Management scenarios

The validated model was used to study the salinity distribution under different management scenarios. A series of model simulations were developed to investigate a number of strategic actions considering natural and anthropogenic salinity sources (Table 6-1).

The SAR system is under increasing pressure due to decreasing water inflows into the river. This is attributed to continuous water developments at the upstream of the SAR. Figure 6-5 shows the decline of the Tigris inflow, the main tributary to the system. The declining rate was approximately 5.4 m^3/s over the 1990-2010 period. In order to investigate the impact of the upstream developments on salinity, driven by tidal forces in the river, ten scenarios were developed (A-J). Scenarios A, B, and C investigate the impact of the river discharge. The major water resources developments in the Tigris basin started in 1989. Therefore, three periods were considered for this study including high flow, normal, and low flow periods. The high flow period was taken as the average discharge in 1988 (200 m^3/s) representing the natural condition before most engineering projects. The average and minimum discharge of the period 1989-2009 was used to represent the normal and low flow periods, 95 and 25 m^3/s respectively. The local salinity pollution caused by wastewater from the agricultural, domestic, and industrial sectors increases salinity especially at midcourse of the river. Several suggestions have been made in the past to divert all drainage water away from the SAR. Also, the main strategic course of action taken by the local authority is to supply all the water demands in the upstream portion of the SAR, and distributing to different users along the river. An artificial canal with a capacity of 40 m^3/s is under construction for that purpose. The other three pairs of scenarios (D-I) investigate the impact of drainage water, water consumption, and the location of drainage water, respectively. The average annual salinity at S6 (2.7 ppt) is specified as the salinity for drainage water, where the major impact from irrigation return flows and wastewater effluents were observed. To investigate the variation of seawater intrusion for any given river discharge, scenario (J) was developed with river discharge ranging from 5 to 120 m^3/s.

Table 6-1. *Simulation scenarios of management actions, considering fixed values for downstream salinity (30 ppt), where L is the distance of seawater intrusion.*

Action	Scen.	Details	Consumption m³/s	Drainage water m³/s	L km
The impact of upstream management on the seawater intrusion	A	River discharge =200 m³/s and upstream salinity=1 ppt	-	-	24
	B	River discharge =95 m³/s and upstream salinity=1 ppt	-	-	28
	C	River discharge =25 m³/s and upstream salinity=1 ppt	-	-	43
	D	River discharge =95 m³/s and upstream salinity=1 ppt	40	-	33
	E	River discharge =95 m³/s and upstream salinity=1 ppt and drainage water salinity= 2.7 ppt	40	20	31
	F	River discharge=45 m³/s and upstream salinity=1 ppt	40	-	80
	G	River discharge =25 m³/s and upstream salinity=1 ppt	40	-	120
	H	River discharge=45 m³/s and upstream salinity=1 ppt, drainage water within the seawater intrusion range (18 km from the mouth)	40	20	80
	I	River discharge=45 m³/s and upstream salinity=1 ppt , drainage water within the seawater intrusion range (60 km from the mouth)	40	20	70
	J	River discharge ranging from 5 to 120 m³/s	-	-	26-80
Managing seasonal variation	K	Seasonal river discharge of 2014, upstream and drainage water salinity represented by measured salinity at S1 and S6, respectively.	40	20	
	L	Same as in K with seasonal river discharge of 1988	40	20	
	M	Same as in K with upstream salinity of minimum 0.25 ppt	40	20	
Barrier construction	N	Barrier construction with no river discharge nor drainage water	-	-	
	O	Same as in N with drainage water	-	20	

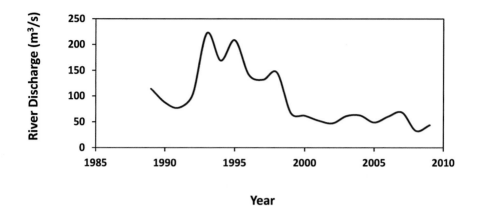

Figure 6-5.*The average annual discharge of the Tigris River at the most downstream gauging station, Missan, (Data collected from local water resources authority).*

Furthermore, the combined seasonal effects of the tidal forces, upstream developments, and local drainage on salinity distribution along the river were investigated. For this purpose another set of model simulations were performed (K, L, and M). These scenarios are to compare the seasonal salinity distribution before and after water resources developments in the basin. Here use is made of the average monthly river discharge of the year 1988 (scenario L) and 2014 (scenario K). The measured salinity at S1 represented salinity of the receiving water for both periods. Drainage water was accumulated at one point near the middle course. The salinity concentration of the drainage water was represented by the measured salinity at station S6.

Local academics and experts have had prolonged debates about the strategic solutions dealing with the escalating problem of the SAR salinity, in which building a barrier across the riverbed is at the top of the list. The barrier is to impound the water and regulate supply primarily for irrigation, domestic, industrial use and power generation. The barrier would have the function to restrain the impacts of seawater on upstream salinity, especially during dry periods or low river discharges, in order to protect the less saline receiving waters in the SAR. The last set of scenarios (N and

O) were developed to study the impact of the barrier construction on the downstream salinity changes.

6.3 Results and discussion

6.3.1 Managing seawater intrusion

Longitudinal salinity distributions for a two-month simulation period during wet (A), normal (B), and dry (C) conditions are illustrated in Figure 6-6. The results show the maximum seawater intrusion length at the end of the simulation period; 24, 28, and 43 km respectively. Large-scale developments obviously change the river flow patterns, reduced river discharge, and extend the seawater intrusion length further upstream. The impact of the upstream developments increased with time (Figure 6-5). Declining quantities of water flowing into the river is considered the main cause of the escalating salinity threats. Recent observations of increase in the amount of seawater intrusion confirm this. This threatens productivity of the estuary ecosystem and large date palm plantations located along its banks.

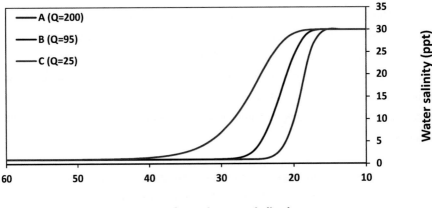

Figure 6-6. *Seawater extent during wet, normal and dry periods.*

To simulate the combined impact of water withdrawals and water drainage on salinity extent, the analysis was run assuming normal conditions. Water withdrawals of 40 m^3/s were specified at the upstream end of the river domain. The simulation applies two conditions of drainage water: firstly, discharging drainage water away from the river course (D) and secondly, discharging drainage water at the downstream portion (E). The volume of drainage water considered is 20 m^3/s, including marsh water, domestic, and industrial effluents (10 m^3/s), and 25% of water consumption (minimum drainage volume according to the Iraqi specification). Figure 6-7a shows that discharging drainage water at the lower portion can serve to reduce the intrusion length rather than drain it out of the system.

The river discharge is seasonally changing. It can be 45 or 25 m^3/s during dry periods. Considering the design capacity of the supplying canal (40 m^3/s), the river discharge could be close to or even less than the water withdrawals along the SAR. To compare the combined effect of water consumption and the changing river discharge, two simulation scenarios were developed to predict salinity changes in both situations, excluding drainage water. Figure 6-7b shows the same pattern of salinity distribution when water withdrawals are equal (F) or more than the river flows (G). The main difference is in the length of seawater excursion, and in seawater being induced deeper as more water is diverted from the system. It can reach up to 80 km in the case of consumption close to the river discharge. Seawater intrusion length can reach up to 110 km when water withdrawal is 15 m^3/s higher than the river discharge. This situation frequently happens during dry periods, so not only decreasing the volume of water inflows has an effect on the seawater intrusion, but also the water withdrawals along the SAR clearly exacerbate the problem. A considerable volume of water is leaving the system during flood periods of tidal cycles through several canals branched from both sides of the river. This, combined with water consumption, allows for further seawater intrusion.

To assess the impact of different effluent points, the location of the outfall of the drainage water was examined. Based on equal water consumption and river discharge scenario F, the impact from drainage water was investigated at two different locations. The first location was chosen in the seawater intrusion domain (H) and the second one at the saline water-freshwater interface point (I). Draining water within the seawater intrusion domain decreased salinity concentration without affecting the intrusion length (Figure 6-7b). Compared to F, salinity concentrations decreased almost by half in the downstream part and then followed the same pattern in the upstream part, while the seawater intrusion distance remains almost the same. The result of I shows that draining water at the seawater intrusion limit reduced the salinity concentrations in the downstream part to the same levels as in the previous one as well as in the upstream part, and also reduced the seawater intrusion length (Figure 6-7b). Therefore, draining water at the saline water-freshwater interface point can maintain both the concentration of salinity and the length of intrusion. Salinity concentrations and seawater extent will be mainly controlled by the water quality and quantity of the drainage water.

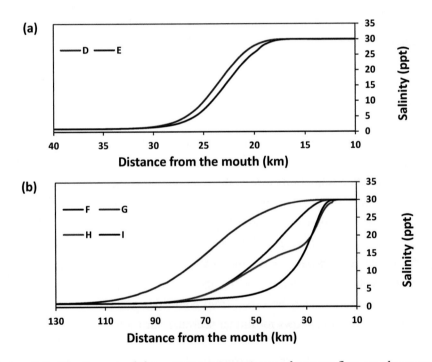

Figure 6-7. *The impact of the water consumption and return flow on the seawater extent, a) water consumption and drainage water under normal conditions of the river discharge, b) changing the river discharge and different location of the drainage water, at 18 and 60 km from the mouth in scenarios H and I respectively.*

The model results show a clear influence of upstream developments on salinity distribution along the estuary. Seawater intrusion distance is extremely affected by flow regulation upstream. Figure 6-8 demonstrates the relationship between the distance of seawater intrusion and river discharge (J), and shows how the distance changes between 26 to 80 km corresponding with a range of river discharge from 5 to 120 m³/s. The Figure presented the non-liner relationship between increasing intrusion distance and decreasing river discharge. The intrusion distance increases significantly with low river discharge, in particular when the river discharge below 40 m³/s, a small change in river discharge contributes to notable increases in the distance of seawater intrusion.

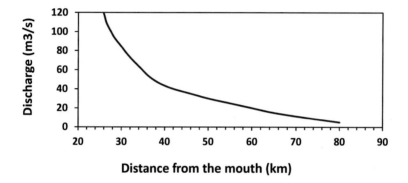

Figure 6-8. *Relationship between river discharge and predicted seawater intrusion length.*

6.3.2 Managing seasonal variation

Figure 6-9 presents the seasonal salinity distribution under changing river discharge at midcourse near the city of Basra. The impact of the upstream salinity and local return flows increased with decreasing water quantities over the simulation periods. The impact of the return flows increased further during the end of 2014 when river discharge was significantly reduced. To compare the improvement of using high river discharge or low salinity concentration on salinity changes, a lowest salinity level (0.25 ppt) recorded at S1 was used together with the river discharge of 2014 (M). Generally, Figure 6-9 shows a clear improvement over the entire year, and salinity found to decrease more during the February-July period where locally, drainage was relatively low, considering the volume of water flowing into the river. High river discharge during 1988 (average flow 200 m^3/s), compared to the year 2014 (55 m^3/s), reduced average salinity near Basra from 2.1 to 1.4 ppt. Using river discharge with average salinity of 0.25 ppt compared to 1 ppt during the year 2014 reduced salinity to 1.7 ppt. Based on the river flow conditions of 2014, increasing river discharge or decreasing its salinity concentrations four times each could reduce the salinity with 34% and 21% respectively.

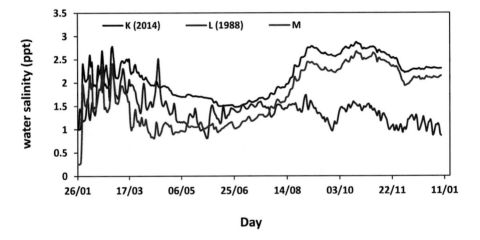

Figure 6-9. *Annual and seasonal impacts on salinity distribution near Basra city in the SAR middle course.*

6.3.3 Impact of barrier construction

The impact of an envisaged barrier structure on the natural sources and water sectors requires detailed studies and design considerations beyond the scope of this limited section. Water resources and the river ecosystem will be affected both upstream and downstream of the barrier. The location of the structure is determined by several factors mainly the extent of seawater intrusion and the shared border between the two neighbouring states. In this study the barrier was assumed to be located wholly in Iraqi territory near the point where the river starts to become the shared river, at 123 km from the mouth. Only its impact on the salinity distribution at the downstream end was investigated.   The salinity variation upstream will be dominated by river flow salinity and the local drainage. Once a barrier is in place it will be necessary to divert all the drainage and wastewater generated upstream to the downstream portion to keep the upstream water as fresh as possible. Therefore, the downstream portion will be under the combined effects of return flows and seawater intrusion. To investigate each effect, two more scenarios were developed. One scenario predicts the salinity variation by considering the drainage water (O) and the other without (N), assuming that there is no flow through the barrier. Figure 6-10 shows the results of both scenarios predicting the salinity changes at the barrier downstream. Figure 6-10a demonstrates the seasonal variation near the confluence with the Karun. In the case of scenario N, the salinity increases continuously over the simulation period. In the case of scenario O, salinity at the lower part will be dominated by the salinity pattern of the drainage water, which is changing based on its quantity and quality. Discharging drainage water at downstream could prevent seawater intrusion further up-river (Figure 6-10b). With the absence of water discharging into the estuary, the seawater intrusion distance could reach up to the barrier location and the salinity concentrations will be dominated by seawater influences. It is expected that the salinity will increase gradually until the river water turns to seawater in the long term. If there is any flow through the barrier, this will improve the conditions in both cases.

6.4 Conclusions

For the first time, the longitudinal salinity variation caused by the combination of salinity sources along the SAR was simulated. The simulation was conducted by using a 1-D hydrodynamic and salt intrusion numerical model.   The model was calibrated and verified using hourly time-series of observed water level and salinity data of the year 2014. Subsequently, the model was used to analyse the effect of different water resources management scenarios on the water and salinity regimes in the SAR.

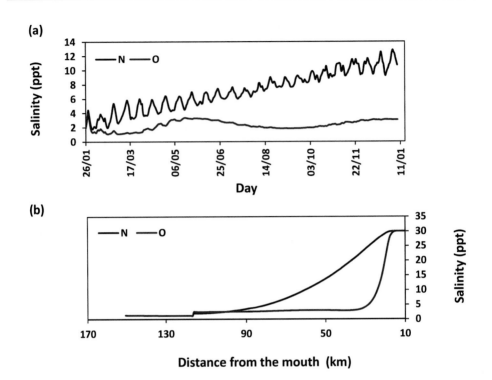

Figure 6-10. *The impact of the water barrier (located at 123 km from the mouth) on the downstream portion with and without drainage water: a) the seasonal salinity changes downstream the Karun at 53 km from the mouth, and b) the longitudinal salinity distribution.*

The analyses were performed with a fixed downstream boundary condition. However, different conditions were specified at the upstream and along the river. The longitudinal salinity distribution was investigated before and after major water developments, considering the tendency of the river discharge, classified into three distinct periods: wet, normal, and dry. The obtained results verify the high correlation between the river discharge and seawater intrusion distance. Human interferences have changed the pattern of river flow and significantly reduced the freshwater discharge, resulting in deeper seawater intrusion landward. The seawater intruded upstream up to 24, 28, and 43 km during the wet, normal, and dry period, respectively. Another set of simulation models were developed to investigate the quantitative changes of flowing into the riverbed considering the water withdrawals and return flows. The model results show significant increases in the seawater intrusion distance and salinity concentration along the estuary. Furthermore, extracting water which exceeded the river discharge allows for more volume of seawater to intrude landward, increasing the seawater intrusion. Discharging drainage water into the river, though with high salinity, could be used to counteract the seawater intrusion. The location of the outfall point affects both the distribution and extent of seawater. The optimal location is determined by several factors such as topography, drainage water quality, and the allowable distance of seawater intrusion.

Salinity changes due to seasonal variation of river discharge was investigated using average monthly river discharges during period of high and low flows, in 1988 and 2014 respectively. The analysis shows that the drainage water combined with the influences of astronomical tides has notably increased the salinity profile during low flow conditions. Managing both upstream water quality and water quantity could mitigate salinity changes along the river but at different levels. Increasing the quantity of upstream releases is more effective in decreasing the salinity concentration along the SAR than improving its quality.

It can be concluded from the results that the salinity variation along the river is a result of combined factors; these are mainly upstream water quantity and quality, local water withdrawals and effluents of agricultural, industrial and domestic wastes, and seawater intrusion. The fact that the river is located in an arid region exacerbates the problems of salinity increments. The results of the hydrodynamic model show a non-linear relationship between seawater intrusion and river discharge, the same trend of salinity as produced by the analytical model (chapter 5). Moreover, the behaviour of salinity dynamic with low flow rate is the same, small change in rate of flow causes notable increases of seawater intrusion distance. Seawater only affects the lower portion, and seawater intrusion could only reach the maximum distance around 80 km corresponding to extremely low river discharge. The difference between the hydrodynamic and analytical models predicting the maximum distance of seawater intrusion under extremely low flow condition, 80 and 92 km respectively, mainly attributed for two reasons: 1) the analytical model did not consider the shallow reach between 40 and 50 km (see Figure 5-3), this could be the main reason for highest estimation of the intrusion distance; and 2) the results of the hydrodynamic model based on fixed value of salinity (30 ppt) specified as downstream boundary condition, while with low river discharge the downstream salinity is higher and could reach up to 40 ppt, therefore the model underestimated the maximum intrusion distance.

The salinity of the upper portion is dominated by upstream influences and local conditions. Therefore, the construction of a barrier at a location further upstream than the maximum seawater intrusion distance will not improve the salinity conditions upstream. Immediately downstream of such barrier, the salinity will be entirely dominated by seawater influences with serious implications on water users and the aquatic ecosystem. The model presented in this study can support the ongoing management efforts describing the salinity variation along the SAR. The availability of further measured data of each salinity source including salinity concentration, discharges, as well as flow direction and velocity could improve the predicting skill of the model.

7 IMPACTS OF DRAINAGE WATER AND TIDAL INFLUENCE ON THE WATER ALLOCATION STRATEGIES *

7.1 Introduction

Improving water allocation strategies for different water uses among different stakeholders is crucial for social and economic developments. Designing such strategies for multiple purposes is complex if water scarcity conditions are prevalent. Distribution of river water whether for irrigation, industrial or domestic use, requires a water quality standard. Seawater intrusion and increasing agricultural and domestic water demands result in salinity increase in many tidal rivers. Managing freshwater inflows is one of the important control measures to maintain salinity fluctuations caused by saline intrusion into river estuaries to a minimum. On the other hand, return flows from irrigation, industrial and domestic effluents increase salinity at their points of discharge and beyond in the downstream direction.

Various methods are now available to model water allocation in river basins. These methods lead to models which are classified into two categories: optimization and simulation models. The determination of an allocation policy is not trivial due to inflow uncertainty and conflicting demands on scarce water resources. Optimization techniques have the ability to deal with multiple conflicting objectives in complex water systems.

Water resources optimization models have been used to determine optimal water allocations among competing water uses. However, little is known about salinity management in general, apart from case specific situations, when there is a

* This chapter is based on Abdullah A., Castro-Gama M.E., Popescu I., van der Zaag P., Karim U., Al Suhail Q., submitted. Optimization of water allocation in the Shatt al-Arab River when taking into account salinity variations under tidal influence. Submitted to *Hydroloical Sciences Journal*.

combination of different salinity sources in a tidal river. There are few studies but these are limited in their scope and applicability and not to the general multi-factors type problems of salinity in a tidal river affected by varied combinations of irrigation practices, industrial effluents and urban discharges (quantity and type), the quality and quantity of upstream river inflow, and seawater intrusion. The information on these factors provides the scientific basis needed to explore effective measures for controlling water quality and resources management. Therefore, there is need for the inclusion of human and natural resources impacts of salinization due to wastewater return flows.

Water allocations in a basin are distributed mainly on the basis of the water system characteristics, water demands and water availability. The objective of this chapter is to develop a multi-objective optimization-simulation model to identify efficient alternatives for water allocation of the SAR. This is necessary to mitigate the impact of high salinity concentrations and fluctuations associated with the different sources on the water utilization along the river. The model is to be used to account for different upstream inflows and salinity concentrations. Furthermore, optimal water allocations and strategies for different upstream conditions are investigated for irrigation and domestic use. For the task set in this study, namely to minimize salinization as well as minimize the deficit for irrigation and domestic water supply in a river system, a hydrodynamic salt intrusion simulation model is required to evaluate the optimal water allocation strategies considering salinity variation under tidal influence.

7.2 Model development

A multi-objective optimization model was coupled to a 1-D numerical simulation model to optimize water allocations in the SAR. Seawater intrusion, return flows, and upstream salinity are accounted to obtain optimized solutions to this class of problems in salinity dynamics. Simulations of different water management scenarios can be introduced individually or in combination to solve for scenarios leading to minimum impact from salt transport. The following section describes this approach.

7.2.1 The simulation model

The 1-D hydrodynamic and salt intrusion model was used to simulate salinity variation and evaluate the objective functions. For this study the period of 26 January to 1 April was specified as the total time interval, which is reasonable for a management objective. The boundary conditions during this interval were the river discharge at the upstream boundary and the hourly measured tidal elevation at the downstream boundary. Salinity at the upstream boundary was specified by the minimum and maximum observed river salinity, 0.25 and 1.5 ppt respectively. At the downstream boundary salinity was averaged over the period and is 30 ppt. The water and salinity levels at the beginning of the simulation period were specified as the initial conditions. The model was developed with 56 observation points to extract the computed salinity (Figure 7-1).

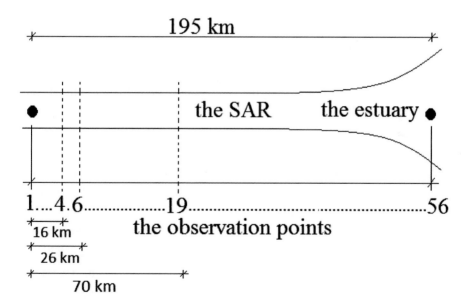

Figure 7-1. *Schematic presentation of the observation points 1-56 showing domestic withdrawals (4), irrigation withdrawals locations (6), and drainage outfall (19) locations.*

Drainage water consists of irrigation return flows and industrial and domestic effluents discharging into the SAR, mainly along its upper portion. Observations in the field and data analysis showed that the maximum impact of the drainage water was at the station near the city of Basra, located at the midcourse of the river (Figure 1-3). The drainage water was represented by a discharge point near Basra, corresponds to observation point number 19 (Figure 7-1). The average observed salinity at this station was specified as the drainage water salinity (2.7 ppt). The rate of drainage water was estimated at 20 m^3/s.

Water withdrawals for both domestic and irrigation uses were represented by two extraction points at distances of 16 and 26 km from the upstream boundary, located at observation points 4 and 6 respectively (Figure 7-1). These are the locations suggested by the local water resources authority to divert water for domestic and irrigation consumption.

7.2.2 The optimization model

The optimization technique was used to establish a tradeoff between the conflicting objectives of minimizing the water supply deficit and minimum salinity. This study aimed at managing the salinity variation along the SAR taking into account growing water demands and decreasing water inflows. Considering the drainage water components, salinity could be mitigated by releasing more water that would dilute and flush out saline water and also counteract seawater intrusion. Discharging more freshwater into the estuary deprive other sectors of their source of water. On the other hand, the salinization impact of drainage water could be extended further upstream driven by tidal influence. Salinity in the upstream portion will be under combined effects: the upstream conditions of water quality and quantity and the downstream conditions of drainage water and seawater intrusion. Therefore, the optimization framework was developed as a multi-objective approach to reduce the impact of salinity changes on the ecosystem and water sectors (domestic and irrigation) under different prevailing conditions. The Monte Carlo technique was used as a computational algorithm for the optimization process coupled with the Delft3d simulation model. The Monte Carlo method is a computational algorithm mainly used for optimization, numerical integration, and probability distribution. The Monte Carlo optimization technique is used for mathematical programming problems, especially nonlinear systems of equations (Dickman and Gilman, 1989), and in optimization (Sakalauskas, 2000 and Guda $et~al.$, 2001).

Two objective functions were considered. The first objective was to minimize the river water salinity concentrations along its course. The second objective constituted minimizing the deficit of water supply (for irrigation and domestic uses). The optimization problem can be mathematically posed as:

$$\text{Min } (F_1, F_2) = [S_{ave}, D_{water~supply}] \tag{7.1}$$

$$\text{Salinity objective } (F_1) = min~(S_i - S_{max}) \qquad i \in \{1, 2, ..., x\} \tag{7.2}$$

$$\text{Supply objective } (F_2) = min~[(A_i - DD) + (A_i - ID)] \quad i \in \{1, 2, ..., x\} \tag{7.3}$$

Subject to

$x \in X;$

where S_{ave} is the average salinity concentration along the river excluding its estuary; the salinity is averaged over the river stretch between observation points number 1 (the upstream end) and number 39 (at 65 km from the river mouth). This is to avoid high salinity levels caused by seawater intrusion as well as to maintain seawater intrusion at that limit, D is the deficit of water supply from two main water uses, irrigation and domestic, and S_{max} is the maximum allowable salinity concentration. The vector of decision variables x lies in the feasible space X determined from the equity and inequity constraints (C) represented mathematically as:

$$0 \leq ID \leq 50$$

$0 \leq DD \leq 10$

$0 \leq S \leq 2.1$

where the maximum irrigation demand (*ID*) is 50 m³/s, maximum domestic demand (*DD*) is 10 m³/s, and *S* for salinity concentration. Water is not allowed for use for consumption if a salinity concentration exceeds the target limit of 2.1 ppt (De Voogt *et al.* 2000). For this reason an additional constraint is posed during the simulation in order to verify whether or not this condition is fulfilled.

The maximum irrigation and domestic demand was estimated considering the available agricultural lands and population of the region, based on the collected information from the competent authorities. The decision variable for the simulation model is the amount of water, *A*, to be allocated from the supply source to each water use, denoted A_i, where *i* designate the user. Available water is a function of the supply source, the season and the development scale. Although water availability varies from season to season and year to year depending on climatologic conditions, the river discharge was expressed equal to the low, average, and high observed monthly discharge over the last two decades. This is to find the optimal water allocation for different uses for each river flow condition.

7.2.3 Coupling of optimization and simulation models

The multi-objective model, written in MATLAB, links optimization with simulation. Figure 7-2 presents a framework of the general algorithm. This code contains a rewritten subroutine within the simulation part which contains the decision variables. The decision variables are generated randomly in the optimization code. A multi-objective optimization routine runs the simulation model for each decision variable. Salinity dynamics is simulated by the Delft3D model. Consequently, the profile of salinity along the river is analysed to evaluate the objective functions and some constraints, the maximum demands and maximum salinity levels. The optimization-simulation model was developed under different scenarios, in order to investigate the impact of salinity changes corresponding to different salinity and inflow levels in the upstream reach of the SAR. To assess the sensitivity of the salinity regime in the river to the drainage water considering different upstream conditions, six scenarios were developed using high, average, and low river discharge, and maximum and minimum upstream salinity (see Table 7-1).

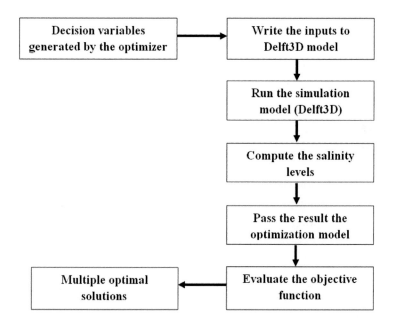

Figure 7-2. *General framework of the multi-objective optimization-simulation model.*

7.3 Results

7.3.1 Simulation-based optimization

Figure 7-3 presents results of salinity distributions at two withdrawals points (for domestic and irrigation supply) and one outfall point (for drainage water). The optimization was used to minimize the average salinity concentration along the river. Generally, the average salinity increases in all scenarios but with different levels. Contrary to expectation, the salinity is found to increase more in the case of minimum upstream salinity (0.25 ppt), not the maximum case (1.5 ppt). Under minimum salinity, S_{ave} increases two, three, and seven-times with high, average, and low flow respectively, whereas, under maximum salinity, S_{ave} increases by 10 %, 30 %, and 60 % with high, average, and low flow respectively. The different salinity range under different upstream conditions could be attributed to dilution and mixing processes along the river.

Table 7-1. *Simulated scenarios (Sc) of salt concentration for different upstream conditions.*

Sc			Salinity of upstream inflow (ppt)	
			0.25	1.5
			1	2
Upstream river	100	**1**	Sc_{1-1}	Sc_{1-2}
discharge	50	**2**	Sc_{2-1}	Sc_{2-2}
(m³/s)	25	**3**	Sc_{3-1}	Sc_{3-2}

Upstream conditions are more dominant at the withdrawal points during high and average flow scenarios (Sc_{1-1} - Sc_{2-2}). Despite increases in S_{ave}, salinity concentration remains almost constant at both domestic and irrigation withdrawal points with slight increases during an average flow. At the outfall point, however, salinity increases, but differently, during high and average flow in both minimum (0.25 ppt) and maximum (1.5 ppt) upstream salinity conditions. In the low flow scenarios (Sc_{3-1} and Sc_{3-2}) the differences are significant. S_{ave} at domestic, irrigation, and outfall points could exceed the allowable limit (2.1 ppt). The red domain in Figure 7-3 represents the salinity concentration where river water is not usable. The red domain increases in downstream direction and extends further upstream with higher upstream salinity concentration.

Figure 7-4 shows the salinity distribution at the withdrawal and outfall points as a function of deficit in water supply. The results show that water demands can be

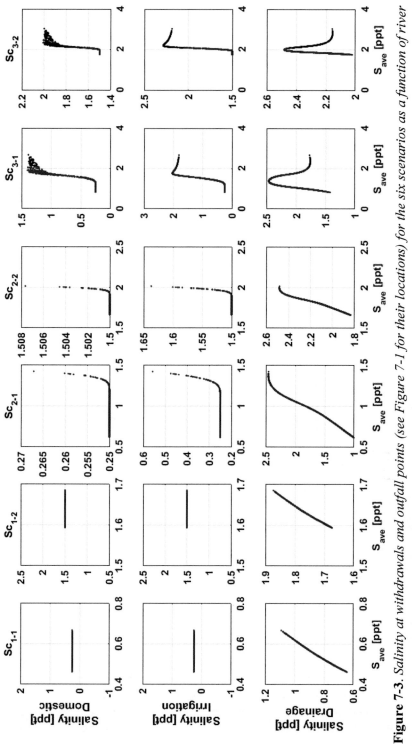

Figure 7-3. *Salinity at withdrawals and outfall points (see Figure 7-1 for their locations) for the six scenarios as a function of river water salinity (note that the axes have different scales for the different conditions).*

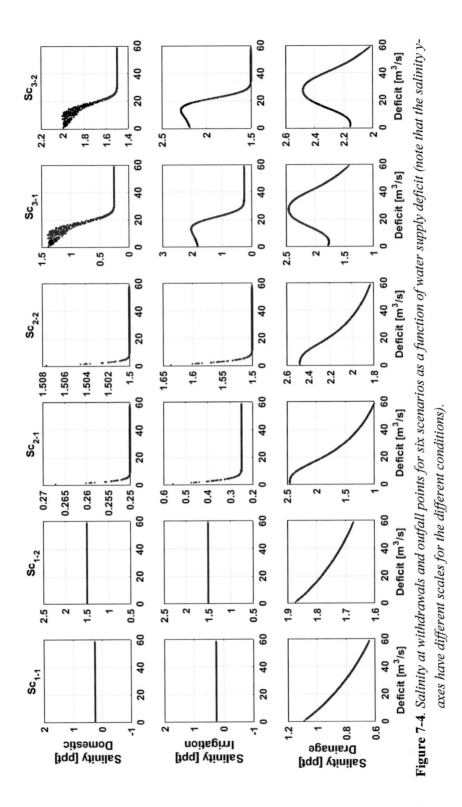

Figure 7-4. *Salinity at withdrawals and outfall points for six scenarios as a function of water supply deficit (note that the salinity y-axes have different scales for the different conditions).*

satisfied during high flow conditions. Drainage water combined with tidal influences has no impact on the salinity levels in the upstream region. However, during average flow scenarios, drainage water could only increase salinity concentration in the upstream region with full water supply, while reduced water supply to domestic and irrigation uses by about 10 m^3/s (16%) upstream salinity levels remain constant. During low flow conditions, there is no possibility to satisfy all water demands, with deficits of around 7 m^3/s (12%) and 20 m^3/s (33%) in scenarios Sc$_{3-1}$ and Sc$_{3-2}$, respectively. Only if the supply deficit is at least 50 % there will be no increases in upstream salinity levels. On the other hand, salinity concentration at the outfall point increases with increasing water supply in all scenarios (Figure 7-4). During high flow conditions the Save rise to 1.1 and 1.9 ppt with minimum and maximum salinity conditions, while it goes up to 2.5 ppt in the other scenarios.

The concentration ensemble of the simulations is presented in Figure 7-5. Drainage water influences the salinity in the downstream reaches under high river discharge. However, its impact extended to the upstream portion with average river discharge and increases further with low river discharge. With minimum salinity levels (Sc$_{1-1}$, Sc$_{2-1}$, and Sc$_{3-1}$), the salinity regions in the river varies more widely compared to maximum salinity levels (Sc$_{1-2}$, Sc$_{2-2}$, and Sc$_{3-2}$). Seawater intrusion, on the other hand, varies according to river discharge and water supply. Consistent with expectation, the seawater intrudes further upstream in case of low river discharge and high water withdrawals.

7.3.2 The optimal solution

In a multi objective optimization problem one cannot get a single optimal solution fulfilling both objective functions simultaneously. Therefore, the aim of multi objective optimizations is not to search for the best solution but to find out an efficient solution that minimizes both objective functions. In fact there are infinite numbers of possible solution points (called the Pareto efficiency). A graphical presentation of the two objective functions in the selected criterion space is presented in Figure 7-6. The feasible ranges of the optimal solutions are bounded by the deficit of water supply at the upper left corner and the average river water salinity at the lower right corner, excluding the red part (Figure 7-6). Considering water supply requirements, the red part indicates that the river salinity exceed the maximum level for irrigation and domestic uses. Each corner represents the optimal benefit of its corresponding individual objective, while all other individual objectives are worse off. Moving from the lower right to upper left points on the convex efficiency front gives different weight to the objective functions, which represent the tradeoff between water demand and salinity level.

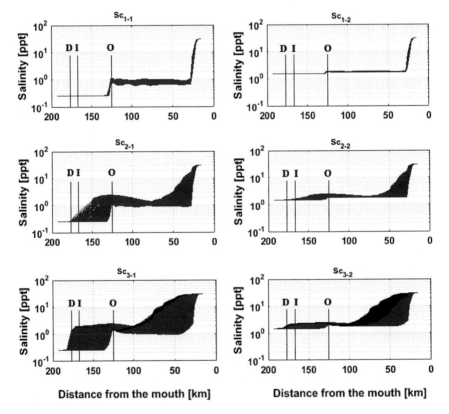

Figure 7-5. *Optimized salinity distribution along the SAR extracted from the 56 observation points, point D, I, and O represent the locations of the withdrawals for domestic water supply, irrigation water supply, and the outfall for drainage water respectively (in red unfeasible solutions, i.e. average river water salinity is larger than 2.1 ppt).*

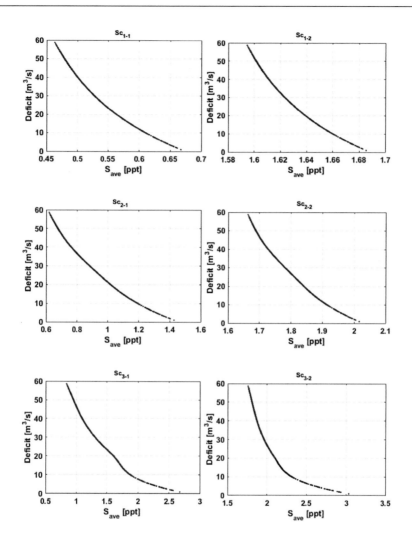

Figure 7-6._The objective space of solutions for each of the six selected scenarios simulated 1000 times (in red unfeasible solutions, i.e. average river water salinity is larger than 2.1 ppt)._

7.3.3 System performance

To evaluate the water system performance under different upstream conditions, the results of one optimal solution for each scenario are presented in Figure 7-7. For that purpose, the middle point of the efficiency front has been chosen that gives equal weight to both objective functions. The Figure shows that the salinity region in the river is influenced by the combined effect of drainage water and seawater intrusion. Tidal influence on the river water salinity extended further upstream with minimum salinity levels under average and low flow conditions (Sc_{2-1} and Sc_{3-1}), whereas it has almost the same extent with high river discharge. Therefore, all seawater intrusion distances are almost the same in scenarios Sc_{1-1} and Sc_{1-2}, around 30 km. During average flow, the seawater intrusion distance corresponding to minimum and maximum salinity levels, increases from 40 km to 60 km, and during low flow increases from 80 km to 90 km, respectively. Out of the seawater intrusion domain, the highest salinity concentrations are at the outfall location but this reduced in both upstream and downstream directions.

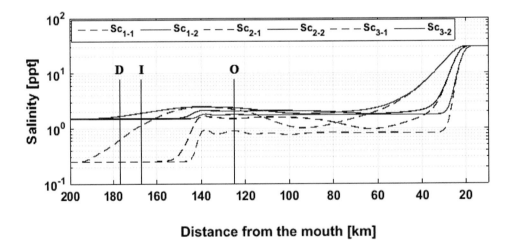

Figure 7-7. *Longitudinal salinity distribution considering different upstream conditions, points D, I, and O represent the locations of the withdrawals for domestic water supply, irrigation water supply, and the outfall for drainage water respectively.*

7.4 Conclusions and recommendations

A multi-objective optimization model was developed incorporating simulation and to determine the impact of drainage water on the natural sources and water sectors under different conditions. This approach was used to address the issue of river salinity changes considering the upstream salinity sources and the tidal influence downstream. Moreover, the water system performance was evaluated under different upstream conditions taking into account demand requirements and salinity variation. This is to assist the decision making process for water resources management and development.

A Monte Carlo optimization algorithm was coupled to a Delft3D simulation model within a MATLAB environment. The optimizer code consisted of two objective functions to minimize risk in water supply and manage the river water salinity to remain in the usable range (lower than 2.1 ppt). The simulation model was used to compute salinity results taking in account the quantity and quality of river and drainage water, water demands, and seawater intrusion. The results were then evaluated by the optimization code to produce an efficiency front of optimal solutions.

Limits of water availability for human uses were obtained in terms of their salinity. The model has been applied to the SAR case. Drainage into the SAR of saline water influences the salinity regime in the downstream and upstream reaches most significantly during low river flow. The range of water salinity variation is mainly governed by the salinity levels of the receiving water, where the impact of drainage water increases at lower salinity levels. Moreover, the impact of seawater extended further upstream at low salinity levels of river water compared to the higher levels. Water demands can only be satisfied under high and average river discharge with salinity concentrations in the range 0.45 to 2.1 ppt. However, under low flow conditions not all water demands can be met, a deficit of 12% and 33% corresponding to minimum and maximum salinity levels of the upstream source.

The presented approach can be used to explore and capture the impact of salinization sources on water sectors and the ecosystem, and is intended to provide an effective tool supporting water resources planning and management. Ideally, this is achieved by conserving computation time using a 1-D hydrodynamic model, reasonably sized grid and time intervals with sufficient resolution, as well as using the simplest Monte Carlo optimization algorithm. To help find more complex patterns for sustainable management of water resources, the existing model can be extended to include more dynamic conditions. For that purpose further work is required considering dynamic salinization sources such as varying drainage volumes and/or salinity levels, and several outfall points for drainage water. This could guide the decision process determining the optimal location of drainage water outfalls while minimizing river salinity and seawater intrusion.

8 SYNTHESIS, CONCLUSION, AND RECOMMENDATIONS

8.1 Water availability and demand

The Tigris, Euphrates, Karkheh, and Karun rivers, which form the Shatt al-Arab estuary, contribute to meeting water demands in Turkey, Syria, Iran, and Iraq. The Tigris and Euphrates Rivers are the primary water sources in Iraq. Historically, the rivers are considered pivotal for the prosperity of ancient civilizations in the fertile Mesopotamian plains. Before the major regulation works on the upstream rivers, the average discharge of the SAR into the Gulf was around 1,600 m^3/s. The major tributaries contribution 13.5, 7.1, 6.3, and 24.7 $\times 10^9$ m^3/yr from the Euphrates (Thi-Qar station), Tigris (Missan station), Karkheh (Jelogir station), and Karun (Ahvaz station), respectively.

In the last three decades, the rivers which feed the delta system are rapidly drying up. Water withdrawals have also increased over time corresponding with population growth and consequent expansions of agricultural and industrial sectors. The flow regime of the system has dramatically changed over time since the construction of numerous large dams for irrigated agricultural and hydropower generation. The inter-annual flow variations of the major tributaries indicating significant declining trends are mostly due to massive developments, annually reducing the natural flow by the following values: the Euphrates 0.6 %, the Tigris 0.3%, the Karkheh 2.0 %, the Karun 0.6 %.

Water flow into the nearby marshes has been altered and inflows into the SAR from this source have largely decreased. The Mesopotamian marshes, which played a central role in the economic and social advancement of around 500,000 people, are considerably affected in terms of water quantity and quality. The area was almost entirely dried out by 1995 and converted to desert lands due to water diversion, over extraction and geo-political reasons. The Tigris is currently the main upstream source of freshwater of the SAR, while the SAR still receives water from the Karun further downstream. The mean annual river discharge into the SAR was reduced to 97 m^3/s, which is only 6 % of the natural flow.

Water scarcity is a most serious problem in the region. The ongoing water resource management and development policies of the riparian countries are most often developed unilaterally. This will put the already stressed system under deeper crisis due to aggravated water scarcity. Discharge may reach close to zero in the future if all planned developments in the upstream rivers are implemented. The demands for Euphrates water in Turkey and Syria can be met, but with serious impacts for Iraq, which will be confronted with significant water shortages. The situation on the Tigris is better, since water availability is so far sufficient to meet the demands. Similar trends are anticipated as a result of current and ongoing water resources developments in the Karkheh and Karun rivers in Iran, as well as the tributaries shared between Iraq and Iran. Estimates of future water demand under full development scenarios for the Euphrates and Tigris rivers, taking into account expected population growth and large-scale developments, respectively in Turkey, Syria, Iran and, Iraq are 32.6, 8.6, 6.2, and 97.5 $\times 10^9$ m^3/yr, almost 1.8 times of the Euphrates-Tigris flows. The total deficit to the SAR is estimated at 53.4$\times$ 10^9 m^3/yr considering both future water demands and investments by the 2040. The water shortages in Iraq will increase which affect all major water uses. The water flows to the SAR system will continue to decrease over time, which will further deepen the ongoing water crisis in the region.

8.2 Water quality deterioration and the consequent impacts

Pollution from agricultural, industrial, and domestic sources has negatively impacted the water quality of the SAR. The available, though limited and fragmented data, revealed a significant deterioration of water quality. The reported increases in the concentration of SO$_4$, Mg, Cl, Na, and Ca clearly indicate high pollution loads and high salinity levels in particular.

Salinity of the major tributaries increased over time. The decreases in river runoff combined with accumulated salt concentration due to increased evaporation and return flows from different activities, led to the increased salinity of the Euphrates and Tigris water; by 2002 salinity levels near the confluence with the SAR were observed at 3.5 and 1.8 ppt, respectively. The salinity of the Karkheh River almost doubled over the last 14 years. The Karun salinity exhibited large inter-annual variations, in the range of 1.0-4.0 ppt during the 1970-2001 period. The deterioration of the Mesopotamian marshes has a significant impact on the SAR. Water salinity of those marshes has increased probably due to low inflows, increased pollution, and high evaporation losses. The discharges of highly saline water, especially from the Hammar and Central marshes, contributed to the observed increases in the salinity levels of the SAR.

The reviewed literature provides an overview of the deteriorating water quality of the water bodies including the SAR but do not provide a consistent set of observations, both in time and space, required for comprehensive situation analysis and thereafter developing management strategies. A salinity monitoring network of 10 stations was installed along the SAR under this research to fill the data gaps. The data from those stations was collected at hourly time step during an entire year in 2014, which provided necessary information to examine the dynamics of this deltaic river system. This unique dataset shows that the salinity of the SAR is high and extremely variable both in space and time. The difference between upstream and downstream salinity levels is significant. The salinity values varied between 0.2 and 40 ppt along the river during 2014. The mean annual salinity at the upstream end was around 1 ppt, increasing to 2.2 ppt at Makel station, near the middle course. Immediately downstream, the salinity increased further at Basra station to 2.7 ppt on average. Further downstream, salinity levels increased and decreased during the year due to the combined impact of the dilution by the Karun waters and incremental impact of seawater intrusion. The highest mean annual salinity level was around 17 ppt in the estuary. Similar to the major tributaries, the annual salinity levels indicate an increasing trend along the SAR course toward the mouth.

However, at monthly and finer time scales, a mixed pattern of increases and decreases was observed along the river course. Based on the observed salinity dynamics, the river could be divided into four distinct reaches, R1-R4 from upstream to downstream: reaches R1 Tigris-Shafi (stations S1-S4), R2 Makel-Abu Flus (S5-S7), R3 Sehan-Dweeb (S8-S9), and R4 at Faw (S10). The ranges of monthly mean salinity concentrations for R1-R4 were 1-2, 2-5, 1-12, and 8-31 ppt, respectively. These changes were attributed to multiple governing factors. The relative impact of these factors varies along the river course, which should be taken into account while developing management actions.

The major factors of increasing water salinity were identified as: salt enrichment in reservoirs and marshes due to evaporation in particular in hot and arid environments; the diversion of more water from the rivers upstream deprives the wetlands of water and reduces the inflow into the SAR; water supply contamination by the surface and subsurface drainage from domestic, industrial, and agricultural fields having high levels of ions and organic materials; local pollution caused by discharging untreated wastewater from different sectors; and the salts entering the river due to the natural tidal phenomenon from the Gulf with evidence of increased seawater intrusion attributed to anthropogenic changes in the river flow regime.

The region is facing a serious problem of increased water salinity for human uses which have detrimental impacts when used for drinking and other domestic purposes. Moreover, salinity increase is considered the main reason of the decline of water productivity in the agriculture sector. Additionally, salinity causes serious economic losses and has notably decreased the fertile land along the river banks, resulting in abandoned farmlands, loss of livelihoods and mass migration.

8.3 Salinity dynamics

The combination of tidal and river discharge fluctuations makes it difficult to recognize the real extent of seawater intrusion and its impact on the longitudinal salinity pattern along the river under different conditions. The analytical approach developed using the field data collected in this study is capable of describing the extent of seawater intrusion along the estuary. In order to utilize the predictive equation for the dispersion, the freshwater discharge into the estuary has to be known. However, it is difficult to accurately determine discharge in the tidal region. The measured river discharge into the estuary was revised considering water withdrawals along the tidal domain, which considerably helped improve the performance of the analytical model. The 1-D analytical model (developed by Savenije 1986) combined with the predictive equation is a quick, effective tool to investigate the impact of seawater intrusion among other sources on the river salinity, and to analyse the dynamics of the saline water-freshwater interface, thereby having the potential to inform water management decision making.

The complex pattern of salinity changes along the river is due to the combined effect of marine and terrestrial salinity sources. Understanding the salinity changes driven by natural and anthropogenic sources is essential to provide the basic step for predicting salinity dynamics and better water management strategies. The longitudinal salinity distribution is highly dependent and controlled by freshwater discharges and tidal forces. The high spatial and temporal gradients require numerical model support to simulate such a complex water system. The hydrodynamic salt intrusion model was pivotal in better understanding the salinity dynamics and has become a norm to test various management scenarios, which otherwise could not be well investigated on the basis of observed data alone.

The 1-D hydrodynamic (Delft3D) model allows the investigation of the salinity changes along the horizontal axis of the entire river, as the transport mechanisms are dominant in the longitudinal directions. The model could also simulate daily and seasonal salinity variations, though the time step was minutes. The model was developed for the SAR with the purpose of informing the decision making process and optimization studies. The model is used to simulate water flows and salinity concentrations along the SAR taking into account the tidal effect of the Gulf at the mouth of the river, and can be used to investigate other conditions or water resources management strategies.

A multi-objective optimization-simulation model was developed to identify efficient alternative water allocation strategies in the SAR. This was necessary to mitigate the impact of salinity concentration associated with different sources on the water use activities along the river. The developed approach successfully captured the impact of river salinization due to return flows on human and natural resources. The multi-objective optimization model yielded explicit trade-off information between the objectives. The model was used to identify optimal water allocation strategies for different upstream conditions and with different allocations for irrigation and domestic use.

8.4 Water resources management

The SAR is the most important water source for the people in this water scarce arid region. The river supports agricultural and industrial practices, navigation activities and biodiverse ecosystems. Water is diverted for irrigation purposes mainly for grain production in the upper course and palm forests in the lower course. Several water treatment plants divert water for domestic use along the river. Water in the SAR system is also used for industrial activities, mainly the expanding oil industry, a sector which provides around 90% of Iraq's national budget.

Today, the region faces serious ecological problems and various challenges of water management, including growing demand, increasing salinity, competition for water between upstream and downstream uses, land degradation and consequent poverty and mass exodus towards urban areas. Given the ongoing pattern of water resources developments, the salinity levels are anticipated to increase further in future. The current efforts on salinity management are not enough to adequately address the mounting crisis which is most alarming in the case of the SAR.

Integrated datasets and modelling approach show that salinity dynamics is mainly governed by upstream, local, and downstream conditions. Seawater intrusion is influenced by upstream conditions, such as flow regulation and water withdrawals. The analytical and hydrodynamic models demonstrate a non-linear relationship between river discharge and seawater intrusion. The distance of seawater intrusion is very sensitive to river discharge when the flow is low. The distance can vary between 24 and 92 km during high and extremely low river discharge, respectively. Water withdrawals play a central role in the extent of seawater intrusion further upstream, especially when river discharge is close to or lower than the water demands. The lower reach of the SAR (R3) exhibited the most upstream and downstream influences. This was due to the tidal influence (downstream) and the dominant impact of the Karun River (upstream). In the upper reaches (R1 and R2), the local conditions and upstream influence were found prominent, while tidal influence is dominant in the estuary (R4). Discharging drainage water into the river, though with relatively high salinity, could be used to counteract seawater intrusion. The location of the outfall point affects both the distribution and extent of seawater. The drainage water combined with the influences of astronomical tides has notably increased the salinity profile during low flow conditions. Managing both upstream water quality and quantity could mitigate salinity changes along the river but in different degrees.

Water allocations are mainly based on the water system characteristics, water demands and water availability. Considering the current population and available agricultural land, the total water allocation in the SAR for 2014 was estimated at 1.9 $\times$ 10^9 m^3/yr, of which about 83% is for irrigation and 17% for domestic demands. However, actual water uses are variable and depend on the salinity of the river water. Drainage water of various user sectors increases river salinity and consequently impacts natural and human uses. The human demand can be only satisfied under high and average river discharge. However, under low flow condition the deficit of water supply could reach up to 33%.

8.5 Contribution and innovative aspects

This PhD research embraces an important theme in arid and semiarid regions: salinity dynamics in rivers under tidal influence. The understanding of salinity variation associated with different factors is of high importance from a scientific point of view, but also is relevant for social, economic, and environmental reasons. The study managed to successfully conduct a monitoring campaign in a data scare and highly dynamic estuary under very tight border and local security conditions, which speaks for the novelty and importance of the collected datasets. The research used the available tools such as Delft3D, 1-D analytical model, optimization, and statistical analysis. The application for the Shatt al-Arab system is new and the integration of these methods in the same envelope is also very interesting and brings in the scientific rigour, innovation, and novelty. The novel data sets collected and consequent analysis and decision support tools developed during this PhD research contribute to understanding the salinity regime in the SAR and provide insights into feasible management strategies. Specifically, this research contributes to improving our understanding of salinity processes at finer spatial and temporal scale and water availability situation exhibited in the delta region, which is diverse in the hydrodynamic features. The research is very important from a social point of view. Extending the work and its participants has therefore the capacity to impact the well-being and economic condition of the populations and can stimulate cooperation among the regions of Iraq and the riparian countries. This effect in Iraq is maximized through tighter connection with local academics, experts, and stakeholders. This thesis and associated peer reviewed publications will reach the broader scientific and decision makers in the riparian countries and beyond.

This PhD research addresses some key issues related to the salinity regime in a delta region and water resources, and highlights its importance and use in water resources planning and management. The dissemination of the study results is vital since it may lead to rational water allocation strategies. These strategies help to sustain acceptable salinity levels across the river, and provide a baseline for further research on improving water resources management in the region. An attempt has been made to specifically address various issues. The main innovative aspects and contributions of this research include: (i) combined analysis of possible salinity sources from the upstream, local, and sea side simultaneously; (ii) implications on salinity simulations in a complex system using the Delft3D model, contributing to improve understanding of the salinity changes and their linkages with different factors; (iii) development of a new multi-objective optimization-simulation approach, and (iv) improved knowledge on spatiotemporal variability of salinity and water resources, through novel data sets and application of rigorous state-of-the-art methods.

8.6 Recommendations and future directions

The central and local Iraqi governments are aware of the salinization of surface water sources particularly of the SAR. However, concrete actions to address the issue remain uncoordinated and limited due to several reasons like lack of reliable data and evidence-based recommendations. The ongoing problems in the SAR region are likely to worsen over time, as can be inferred from the patterns observed in water resources development and management in the region. Water availability in the SAR region and Iraq is affected by upstream water developments. To solve the problem that the SAR is currently facing requires concerted efforts both within Iraq and with the upstream riparian countries.

Iraq needs to urgently apply salinity management actions to control discharge of the saline water return flows from the irrigation systems within Iraq. Moreover, wastewater treatment also requires accelerated efforts to avoid salinity increase and avert polluted and saline water entering into the water bodies in Iraq. At the local scale, the discharge of saline water coming from the marshes needs to be controlled and alternative measures of disposal need to be examined, and accumulated salt required to be adequately drained.

The available information on discharge and salinity of some sources is limited and not fully conclusive. Water contributions from the major tributaries including the marshes to the total flow volume into the SAR required specific measurements. Detailed investigation considering water withdrawals and water losses could be valuable for estimating the water balance of the SAR system. Further monitoring is required for the point sources of salinity along the SAR and its main tributaries, including wastewater discharges from the main cities, the main drainage canals from irrigation zones, as well as from the main industries especially from oil production. The monitoring program should also include more water quality parameters besides the flow velocity and direction to estimate water discharges and hence the salt loads caused by different sources. These datasets will provide additional information for improved decision making. Moreover, the additional data along with available sources should be used for improving the analytical, modelling, and optimization tools developed in this research. It is worth exploring new avenues of developing a single and integrated decision support tool with automated database management of datasets from an integrated monitoring network.

SAMENVATTING

Dichtbevolkte gebieden in delta's in warme klimaten zijn kwetsbaar als het gaat om acute beschikbaarheid van water en de waterkwaliteit. Problemen die vaak aan elkaar gerelateerd zijn. Een van de grootste bedreigingen van de waterkwaliteit in deze gebieden is het zoutgehalte. Afnemende aanvoer van zoetwater van slechtere kwaliteit behoren tot de belangrijkste problemen in veel van de deltagebieden en het sociaal-economische / mens- en ecosysteem wat er van afhankelijk is. Vooral in de meest benedenstroomse gebieden dichtbij de grootste zoutbron afkomstig van het zeewater, intensieve landbouw, industriële en bevolkings-hubs. Hoe meer oppervlaktewater wordt gebruikt om de vraag naar zoetwater te vervullen hoe verder het getij en zoutindringing zullen binnendringen. Deze situatie verergerd gedurende aanhoudende periodes van droogte en met afnemende kwaliteit van het water wat terugvloeit in de rivier. Onbetrouwbare waterbeschikbaarheid en achteruitgang van het ecosysteem gaat gepaard met grote sociaal-economische gevolgen. Sommige van de directe oorzaken en gevolgen van zoutindringing zijn direct te zien met het blote oog, maar het volledige beeld op de lange termijn en de strategieën voor verbetering zijn minder duidelijk.

Het proces van verzilting in rivieren is dynamisch en complex. Het is multi-variate en kan zeer variabel/veranderlijke zijn, zelfs willekeurig gegeven de vele factoren en onzekerheden die ermee verbonden zijn. Het bepalen van de factoren is data-intensief en gebiedsgebonden en dient derhalve op maat bepaald te worden voor de in het verleden en huidige tijd heersende omstandigheden met betrekking tot zoutgehalte, hydrologie en watergebruik in de delta. Studies naar het zoutgehalte in complexe delta systemen vereisen derhalve betrouwbare historische data welke gebaseerd zijn op systematische veldmetingen, bij voorkeur met een breed scala aan zoutgehalte metingen. Dit is onmisbaar voor het ontwikkelen van verifieerbare analytische en numerieke modellen welke gebaseerd zijn op gebiedsspecifieke gekalibreerde parameters. Met de juiste monitoring en voldoende verfijnde modellen kunnen rationale diagnostische en beleids beslissingen worden gemaakt voor vermindering van zowel waterschaarste als verzilting. Dit zou uiteindelijk moeten leiden tot duurzaam watergebruik.

Het huidige onderzoek betreft een eerste systematische monitoring en model studie naar de waterbeschikbaarheid, waterkwaliteit en indringing van zoutwater van de Shat al-Arab rivier (SAR). De SAR loopt gedeeltelijk langs de betwiste internationale Irak-Iraanse grens, welke in het verleden een bron van conflicten is geweest. Het is een olierijk gebied waar, sinds het begin van de beschaving, altijd bevolking en economische activiteiten zijn geweest. Het gebied kenmerkt zich door's werelds grootste moerassen en bossen met dadelpalmen. De Tigris en Eufraat rivier aan Irakeze zijde en de in Iran ontspringende Karkheh en Karun rivier, vormen de belangrijkste bronnen van aanvoer voor de rivier en de basis voor de regio's rijke ecologische, sociaal-economische en culturele erfgoed. De monding van de rivier aan de Golf is ook belangrijk als scheepvaartroute waar de dichtbevolkte Iraakse stad Basra met zijn enorme haven opbouwt. Dit onderzoek is actueel en relevant aangezien de problemen met betrekking tot herkomst, regelgeving en beheer van het zoutgehalte in de SAR sterk gepolitiseerd zijn en veel besproken worden op provinciaal en nationaal niveau. De oorzaken en hoogtes van het zoutgehalte worden nog niet goed begrepen, laat staan besproken, wat leidt tot tegenstrijdige percepties over de oorsprong (extern of intern), de natuurlijke condities en de gebruiken die de huidige kritische condities kunnen verklaren. Om de zaken nog ingewikkelder te maken, zijn de rivier toevoeren en grenzen gereguleerd door verschillende provinciale en nationale entiteiten zonder enige coördinatie op regionaal of internationaal niveau. Verbetering en samenwerking door alle partijen op de verschillende niveaus is onwaarschijnlijk gegeven het ontbreken van een gezamenlijk begrip over de oorzaken die hebben geleid tot de huidige situatie.

De huidige wetenschappelijke kennis over de problematiek met het zoutgehalte van de SAR is onvolledig, gedeeltelijk door de complexe, dynamische, ruimtelijke en temporele interactie tussen zoutbronnen, waterontrekkingen en waterafvoeren van de gebruikers uit de verschillende sectoren. Het belangrijkste doel van dit onderzoek is om aan de hand van voldoende juiste en consistente metingen, zowel in plaats als in tijd, de zoutgehalte dynamiek van de delta van de SAR te beschrijven. Dit onderzoek heeft gebruikt gemaakt van verschillende modellen en deze ook gecombineerd, onderbouwd met een netwerk van waterkwaliteitsmetingen (met behulp van divers). Voor het eerst is een systematisch, uitgebreid en nauwkeurig monitoring programma van het zoutgehalte en de waterstand over de gehele lengte van het estuarium en de rivier (200 km) ontwikkeld. Dit heeft geleid tot een unieke dataset in een gebied met extreem zware klimatologische omstandigheden en strenge veiligheidsmaatregelen. Tien diver meetstations zijn op zorgvuldig geselecteerde plaatsen geïnstalleerd, goed onderhouden en regelmatig gekalibreerd. Gedurende heel 2014 zijn elke uur de waterstand, temperatuur en zoutgehalte gemeten. Door de data van 2014 te combineren met historische gegevens van lokale water instanties, heeft dit deel van het onderzoek de statistische, temporele en ruimtelijke verdeling van het zoutgehalte vastgesteld.

Het zoutgehalte is een multivariate probleem in zowel temporele als ruimtelijke dimensies wat een combinatie van methoden en instrumenten vereist vanuit water techniek en management, optimalisatie van systemen en scenario analyses door numeriek modelleren en simulatie methoden. Verschillende model technieken zijn achtereenvolgens gebruikt om data te analyseren en het complexe systeem te simuleren, inclusief kwantitatieve analyse, een analytische methode, een numerieke model en een optimalisatie techniek.

Een kwantitatieve analyse van de 2014 dataset, gecombineerd met de historische data, heeft geresulteerd in een complete beschrijving van de huidige stand van de hydrologie en geografie en laat een ernstige afname van de water hoeveelheid en escalerende waardes van het zoutgehalte zien in de tijd (hoofdstuk 3). De analyse betreft zowel de SAR als de instromende rivieren (Eufraat, Tigris, Karkheh en Karun) met de verbonden moerassen, wat essentieel is om een holistisch beeld te presenteren. De analyses zijn gebaseerd op de meest recente data, hoewel beperkt, de beschikbaarheid van water, ontwikkelingen en infrastructuur met betrekking tot water en toestand van de waterkwaliteit. Wateraanvoeren bleken significant gereduceerd. De toestand van de waterkwaliteit is verslechterd en heeft in 2014 een alarmerend niveau bereikt, in het bijzonder van Basra tot de monding. De oorzaken voor het gestaag toenemende zoutgehalte verschillen per locatie, maar zijn onder andere de afnemende waterkwantiteit en -kwaliteit van de hoofd- en zijtakken, zeewater intrusie door getijde-invloeden, slecht gereguleerde wateronttrekkingen, vervuilde waterafvoeren vanuit de geïrrigeerde landbouw en verschillende andere afvalwater punten, hoge verdamping en af en toe zoutwater lozingen van de omliggende moerassen.

Een analyse van de intra-jaarlijkse variabiliteit van het zoutgehalte laat een grote temporele en ruimtelijke variabiliteit zien in de orde van 0.2 – 40.0 ppt (of g kg-1; x1.5625 µS.cm-1). Overeenkomsten in de dynamiek van het zoutgehalte zijn gebruikt om de rivier in vier afzonderlijke ruimtelijke secties (R1-R4) te verdelen om respectievelijke management maatregelen te beschrijven (hoofdstuk 4). Het gemiddelde zoutgehalte varieert tussen 1.0-2.0, 2.0-5.0, 1.0-12.0 en 8.0-31.0 ppt voor sectie R1 (Qurna tot Shafi), R2 (Makel tot Abu Flus), R3 (Sehan tot Dweeb) en R4 (Faw vlakbij het estuarium), respectievelijk.

Door langs- en vertikale zoutgehalte metingen te correleren konden initiële schattingen van de omvang van de zeewater intrusie in het SAR estuarium worden gemaakt. Om een meer fysisch onderbouwde schatting van de zeewater intrusie te kunnen maken is een voorspellend model ontwikkeld op basis van de specifieke getijde, seizoensgebonden afvoer variatie en geometrische kenmerken van de SAR (hoofdstuk 5). De zeewater excursie is analytisch gesimuleerd met behulp van een 1-D analytisch zoutintrusie model, welke onlangs is aangevuld met een vergelijking voor getijdemenging. Het model is toegepast om de seizoensgebonden variatie van de verdeling van het zoutgehalte te analyseren onder verschillende rivier condities: gedurende natte en droge periodes, spring- en doodtij tussen maart 2014 en januari 2015. Een goede vergelijking tussen de gemodelleerde en geobserveerde verdeling van het zoutgehalte is verkregen. Het schatten van wateronttrekkingen langs het estuarium verbeterde de prestatie van het model, vooral bij de lage afvoeren alsmede met een goed-gekalibreerde dispersie-excursie relatie van de verbeterde vergelijkingen. De lengte van de zeewater intrusie, gegeven de huidige metingen, varieert tussen de 38 en 65 km gedurende het jaar van de metingen. Bij extreem lage rivierafvoeren wordt een maximale intrusielengte van 92 km berekend. Deze nieuwe berekeningen laten zien dat de SAR, reeds geplaagd door extreme verzilting, snel de situatie zal bereiken waarbij ingrijpen ofwel niet effectief meer zal zijn ofwel veel moeilijker en duurder zal zijn. Verschillende scenario's zijn vervolgens onderzocht om dit punt aan te tonen.

Een 1-dimensionaal numeriek hydrodynamisch en zoutindringings model is toegepast om het complexe zoutgehalte regime te simuleren. Het regime is complex door het gecombineerde effect van terrestrische en mariene bronnen (hoofdstuk 6). Het model is wederom gebaseerd op de uur-dataset van 2014. Met het model is de impact van verschillende management scenario's op de zoutgehalte variatie onder verschillende omstandigheden geanalyseerd. De resultaten laten een hoge correlatie zien tussen de variatie in het zoutgehalte en de rivierafvoer. Toename van watergebruik bovenstrooms en toename van lokale wateronttrekking langs de rivier zullen verder bijdragen aan de zeewater intrusie en de toename van zoutconcentraties in de SAR. Het verbeteren van de kwantiteit en kwaliteit van de bovenstroomse water bronnen kan de zoutconcentraties verminderen. Terugvoeren van waterafvoerstromen van menselijke gebruik, alhoewel zout, naar de rivier kan zeewater intrusie tegengaan, gegeven dat de locatie van zulke lozingen zowel de verdeling als de omvang van het zoutgehalte beïnvloedt. De voor de SAR gekalibreerde parameters, numerieke scenario analyses waren bijzonder nuttig om de variatie in zoutgehalte langs de lengteas te bestuderen onder extreme omstandigheden voor elk van de variabelen. Met het aangenomen slechtste scenario kunnen de beste water management strategieën worden beoordeeld, maar dit vereist tevens een trade-off analyse tussen wateronttrekkingen en zoutgehalte.

Een multi-objective optimalisatie-simulatie model is hiervoor ontwikkeld (hoofdstuk 7). Het gecombineerde zoutgehalte systeem, inclusief bovenstroomse zoutbronnen, waterafvoerstromen en zeewater intrusie is gesimuleerd met behulp van een gevalideerd hydrodynamisch model, welke de distributie van het zoutgehalte in de rivier modelleert onder verschillende waterverdeling scenario's. Zes scenario's zijn voorgesteld, getest en gepresenteerd. Het model zoekt naar de optimale oplossing waarbij zowel het zoutgehalte in de rivier als het tekort aan waterlevering voor drinkwater en geïrrigeerde landbouw wordt geminimaliseerd. Het model is gebruikt om de trade-off tussen deze twee doelstellingen te onderzoeken. De ontwikkelde methode, het combineren van een simulatie en optimalisatie model, kan besluitvorming informeren voor een beter beheer en de vermindering van de verzilting in de regio.

De resultaten van de gecombineerde methode met vereenvoudigde veronderstellingen heeft een vrij complex water systeem gereduceerd tot een beheersbaar 1D model. De nieuwe datasets en de daaruit voortvloeiende analyses hebben geresulteerd in een nieuw beslissings-ondersteunings-model, welke met verder aanpassingen, nog ingewikkelder scenario's aankan. Het onderzoek concludeert dat kennis van de prevalerende hoge zoutgehaltes in een complex en dynamisch delta-rivier systeem een centrale rol speelt in het ontwikkelen van maatregelen voor een duurzaam gebruik en beheer van het watersysteem. Deze studie heeft een solide basis gegeven voor de benodigde kennis. De uitgebreide en gedetailleerde dataset vormt de basis van een gevalideerd analytisch model welke de omvang van zee water tussen andere zoutbronnen in een estuarium kan voorspellen. Tevens vormt het de basis voor een hydrodynamisch model dat de veranderingen in zoutgehalte kan voorspellen in een intensief gebruikt en veranderend watersysteem. De aanpassingsmogelijkheden van de modellen aan veranderende omstandigheden maken ze direct toepasbaar voor degene verantwoordelijk voor water management. De methode kan worden toegepast in andere systemen.

De huidige inspanningen bij het beheer van zoutgehaltes zijn niet genoeg om de op handen zijnde crisis adequaat te adresseren. Continue monitoring van waterkwaliteit kan de relatieve impact van de verschillende zoutbronnen op verschillende tijden, vooral zeewater en lokale zoutbronnen, lokaliseren en beoordelen. Bij het managen van zeewater intrusie en elk lokaal effect moet rekening gehouden worden met variatie in kwantiteit en kwaliteit van irrigatie waterafvoeren en afvalwater afvoeren langs de SAR en langs de Eufraat, Tigris, Karkheh en Karin rivieren. De crisis kan enkel worden afgewend door samenwerkingsinitiatieven op het gebied van waterbeheer welke genomen moeten worden door alle oeverstaten. Dit vereist een verschuiving van het huidige model waarbij het waterbeheer unilateraal geregeld wordt naar een internationale samenwerking en beheer van een gedeeld watersysteem. De steun van de regionale en internationale gemeenschappen kan bijdragen aan deze verandering. De op vermindering gerichte crisis strategieën moeten manieren vinden om de aanvoer van bovenstrooms te laten toenemen en de waterkwaliteit te verbeteren. Tegelijkertijd zijn lokale maatregelen nodig om te voorkomen dat slechte kwaliteit huishoudelijk en industrieel afvalwater en het zeer zoute water uit de moerassen in de SAR stromen. Deze inspanningen moeten worden ondersteund door degelijk wetenschappelijke informatie.

SUMMARY IN ARABIC

دلتا اللأنهر الكثيفه بالسكان والواقعه في الأقاليم الجافه تُعتبر الأكثر عرضه لمشاكل المياه كماً ونوعاً ، تلك المشاكل التي غالباً ما تكون مترابطه ومتداخله . واحدةٌ من أكبر التهديدات لنوعية المياه في مثل تلك الأقاليم هي الملوحة . حيثُ أن تناقص الأيرادات المائيه وتدهور نوعيتها من القضايا الرئيسيه التي تتعرض لها الكثير من دلتا الأنهر و النظم البيئيه والبشريه التي تعتمد عليها ، وبشكل ملحوظ في نهايات مجاري الأنهر ، والتي غالباً ما تكون قريبه من مصادر الملوحه الرئيسيه متمثلتاً بمياه البحر والزراعه الكثيفه و مختلف الفعاليات البشريه والصناعيه . حيث إن أغلب المياه السطحيه يتم تحويلها لتلبية الأحتياجات المائيه خصوصاً خلال فترات الجفاف الطويله ، وكثيراً من المياه المستخدمه ما تعاد الى المجرى المائي الرئيسي بنوعيات رديئه ، مما ساهم في زيادة تأثيرموجات المد و نفاذ مياه البحر في المصبات والانهرلمسافاتٍ أطول . إن إنعدام الأمن المائي وتدهور النظم البيئيه له آثار اجتماعية واقتصادية كبيرة . بعض من الأسباب و الأثار المباشرة للملوحة قد تكون واضحه بالعين المجردة ولكن الصورة المتكاملة تكون أقل وضوحاً من حيث العواقب واستراتيجيات الإصلاح خصوصاً على المستوى طويل الأمد .

عملية التملح في الأنهار عمليه ديناميكيه ومعقده. حيث إنها متعددة العوامل والتي غالباً ما تكون متغيره بشكل مستمر وغير محدد . تحديد هذه العوامل يتطلب جمع بيانات مكثفه ولكل حالة معينة ، وبالتالي يجب أن تكون مصممه خصيصاً للظروف الميدانية السائدة في الماضي والحاضر من حيث التراكيز الملحيه و النظام الهيدرولوجي واستخدامات المياه . دراسات الملوحه في نظم الدلتا المعقده تتطلب بيانات تأريخه موثوقه تعتمد على الرصد الميداني المنهجي ويفضل أن تغطي جميع مصادر الملوحه المحتمله . وهذا أمر يُعتبر ضرورياً لبناء النماذج (الموديلات) التحليلية والعددية التي يمكن التحقق منها ومعايرتها بالأعتماد على عدد من المعاملات والتي بدورها تحدد النظام المائي . ان الرصد والنمذجة المناسبه المستخدمه لصقل القرارات التشخيصيه العقلانية وسياسة العمل يمكن توظيفها من أجل تخفيف كل من شحة وملوحة المياه على حد سواء ، والتي ينبغي أن تؤدي في النهاية إلى التنمية المستدامة للمياه .

هذا البحث يمثل عمليه مراقبه حقليه ودراسة نمذجة تعتبر الأولى من نوعها شاملتاً توفر المياه ونوعية المياه ونفاذ مياه البحر في شط العرب . شط العرب الذي يجري جزئياً على طول الحدود الدوليه المتنازع عليها بين العراق و ايران ، والتي كانت مصدراً للصراع في الماضي . تلك المنطقة الغنية بالنفط و التي تتميز بكثرة

السكــان والزراعـة منـذ بدايـة الحضارة، وتتميز ايضاً بوجود أكبـر المستنقعـات (الاهوار) وغابات النخيـل في العالم .

يُعتبر نهري دجلة والفرات في الجانب العراقي ، ونهري الكرخة والكارون النابعة من إيران مصادر المياه الرئيسية لشط العرب . وكذلك مصدراً للتنوع البيئي والأجتماعي والأقتصادي والثقافي في المنطقة . كما أن مصب النهر والمرتبط بالخليج العربي يُمثل طريق الشحن الأهم في المنطقة حيث تقع محافظةُ البصره ، المدينة العراقية ذات الكثافه السكانيه العاليه الغنيه بالحقول النفطيه و التي تتميز بوجود الموانئ الضخمة ومحطات تصدير النفط . البحث الحالي يلبي حاجه ملحه ويربط القضايا ذات الصلة بمنشئها ، حيث ان تنظيم وإدارة الملوحـة في النهر أصبح مُسيساً إلـى حدٍ كبير، وسبباً لناقشـات ساخنـة علـى المستويـات الإقليميـة والوطنية.

أسباب ومستويات تركيز الملوحة لم تكن مفهومة تماماً حتى الأن ، ناهيك عن التعامل معها او معالجتها ، مما يؤدي إلى تضارب المفاهيم حول أسبابها (داخلية أو خارجية) وتأثير كل منها ، والظروف الطبيعية والممارسات التي يمكن أن تفسر الظروف الحرجة الحالية من التراكيز الملحيه العاليه . وما يزيد الأمر تعقيداً ، أن الأطلاقات المائيه والحدود الأداريه يتم تنظيمها من قبل كيانات إقليمية ومحلية مختلفة مع محدودية التنسيق على كلا المستويين الإقليمي و الدولي . ان تعاون جميع الأطراف علـى مختلف المستويات لمعالجة مشكلة الملوحه من غير المرجح في الوقت الراهن خصوصاً في ظل عدم وجود فهم مشترك لمـا تسـبب فـي الـوصول لمثل تلك الظروف الحالية .

ان المعرفة العلمية الحالية لمشكلة الملوحة في شط العرب غير متكاملة ، ويرجع ذلك جزئياً إلى التغير الزماني والمكاني و تعقيد و ديناميكية التفاعل بين مصادر الملوحة ، مضافاً لها الأستهلاكات المائيه و كميات المياة العائده من قبل المستخدمين في قطاعات المياه المختلفة . كان الهدف الرئيسي من هذا البحث هو تقديم مجموعة مدققه ورصينه بما فيه الكفاية من قاعدة بيانات على كلا البعدين المكاني والزماني ، وذلك لدراسة ديناميكيات الملوحة في دلتا شط العرب . واستندت المنهجية المستخدمة في هذه الدراسة على توظيف والجمع بين طرق النمذجة المختلفة ، مدعومتاً بالبيانات الميدانية التي تم جمعها من خلال شبكة من أجهزة الاستشعار لنوعية المياه (مجسات). وقد تم إعداد برنامج رصد منهجي وشامل ودقيق لكل من الملوحة ومنسوب الماء على امتداد كل من المصب و النهر بطول (200 كم) لأول مرة في ظل ظروف أمنية صارمة ومناخية قاسية للغاية ، مما أسفر عن مجموعة بيانات فريدة من نوعها. حيث تم تركيب عشر محطات من المجسات في مواقع تم اختيارها بعناية ، مع الأخذ بنظر الأعتبار استمرار معايرتها وصيانتها بانتظام . حيث تم جمع بيانات لكل ساعه عن مناسيب المياه ودرجات الحرارة والتراكيز الملحية خلال كامل العام (2014) . هذا الجزء من الدراسة مع مجموعة البيانات التي تم جمعها من الأدارات المحلية للمياه لنفس العام قد استخدم لايجاد التوزيع الإحصائي والزماني و المكاني للملوحة والأسباب الرئيسية لها .

الملوحة هي مشكلة متعددة المتغيرات في كل الأبعاد الزمانية والمكانية و التي تتطلب تركيباً من الأساليب والأدوات المستخدمه في هندسة المياه وإدارتها ، وطرق الحل الأمثل ، وتحليل السيناريوهات ، واستخدام

أساليب النمذجة العدديه وطرق المحاكاة . حيث تم استخدام طرق نمذجة مختلفة بشكل متناسق لتحليل البيانات
ولمحاكاة هذا النظام المعقد ، بما في ذلك التحليل الكمي والنمذجه العدديه و التحليليه وطريقة الحل الأمثل .

التحليل الكمي لمجموعة البيانات لعام 2014 جنباً إلى جنب مع البيانات التاريخية ، أسفرت عن وصف موسع
لهيدرولوجية وجيوغرافية النهر في الوقت الحالي والهبوط الحاد في كمية المياه وتصاعد مستويات الملوحة على
مر الزمن (الفصل 3) .تلك التحليلات قد غطت شط العرب وكذلك جميع روافده (الفرات ، دجلة ، الكرخة ،
الكارون) مع الأهوار التي تتربط بها ، والذي يعتبر أمراً ضرورياً لتقديم الصورة الشاملة للتغيرات
الهيدرولوجيه . استندت التحليلات على أحدث البيانات ، وإن كانت محدودة ، فيما يخص وفرة المياه وتنمية
الموارد المائية والبنية التحتية والإدارية و نوعية المياه . والتي تشير الى إنخفاض في كميات المياه وبشكل
ملحوظ . كما أن نوعية المياه قد تدهورت لتصل إلى مستويات عالية وبشكل مقلق حتى عام 2014 ، خاصتاً في
المسافه بين مركز مدينة البصرة و مصب النهر . الأسباب التي يمكن أن تُفسر الزيادة المستمرة في تراكيز
ملوحة المياه تختلف من مكان إلى آخر. والتي تشمل : إنخفاض كمية المياه ونوعيتها (من مصادر المياه الرئيسية
والفرعية) ، و تسرب مياه البحر تحت تأثيرات المد والجزر ، و سوء استخدام المياه وإدارتها ، و تصريف المياه
الملوثة الى النهر (من الري والعديد من نقاط تصريف مياه الصرف الصحي) ، و معدلات التبخر العالية ، وبين
الحين والآخر تصريف المياه المالحة من الاهوار المحيطة .

كما ان تحليل تغيرات التراكيز الملحيه خلال السنه يظهر التباين الزماني و المكاني العالي الذي يمتد من 0.2 و
حتى 40.0 جزء في الألف (غم\كغم) . إن التشابه الذي وجد في ديناميكية الاملاح قد استخدم لتقسيم مجرى
النهر إلى أربع وحدات مكانية مميزة (ر1- ر4) لتسهيل تحليل البيانات مناطقيا والاجراءت الواجب اتخاذها (
الفصل 4) . حيث تراوحت المعدلات الشهريه للتراكيز الملحيه حسب المناطق بين 1.0-2.0 ، 5.0-2.0 ، 1.0-
12.0 , 8.0-31.0 غم\كغم والتي لوحظت على أمتداد ر1 (القرنه - الشافي) و ر2 (المعقل - ابوفلوس) و ر3
(سيحان-الدويب) و ر4 (الفاو قريباً لمصب النهر) على التوالي .

قياسات التراكيز الملحية في كلا المحورين الطولي والعمودي قدمت التقديرات الأولية لمسافة إمتداد مياه البحر
باتجاه المصب وأعلى مجرى النهر . ولغرض تحقيق تقدير أدق لمسافة نفاذ مياه البحر على أسس فيزيائية ، فقد
تم تطوير موديل للتنبؤ بامتداد اللسان الملحي آخذاً في الأعتبار خصائص المد والجزر و التقلبات الموسمية
وتغيرات التصاريف والخصائص الطوبغرافيه للنهر (الفصل 5) . حيث تم محاكاة نفاذ مياه البحر تحليلياً
باستخدام موديل تحليلي لنفاذ المياه المالحه احادي البعد مع المعادلات التي تم تحديثها مؤخراً والمتعلقه بالتشتت
الملحي تحت تأثير المد والجزر . تم تطبيق الموديل في ظل ظروف هيدرولوجيه مختلفة للنهر لتحليل التباين
الموسمي لتوزيع الملوحة خلال الفترات الرطبه والجافة أثناء المد الاقصى والمد المحاقي في الفتره بين آذار
2014 وكانون الثاني 2015 . أضهرت النتائج تطابق جيد بين التراكيز الملحيه المحسوبه والبيانات الحقليه . كما
ان تقدير كميات المياه المسحوبه على امتداد النهر لغرض الاستهلاك اليومي قد حسن من أداء الموديل
وخصوصا في فترات الجريان المنخفض والمعايره الجيده لمعادلات التشتت الملحي . أطوال مسافة نفاذ مياه
البحرحسب القياسات الحقليه تراوحت بين 38-65 كم خلال فترة الدراسه . في التصاريف المائيه المنخفضة

للغاية من المتوقع أن تبلغ المسافة حوالي 92 كم . تظهر هذه التوقعات الجديدة بأن النهر يعاني بالفعل من الملوحة الشديدة ، و يقترب سريعاً من الحالة التي سيكون فيها التدخل إما غير فعال أو أكثر صعوبه و أكثر كلفة . لاحقاً تم دراسة عدد من السيناريوهات لإثبات هذه النقطة .

أيضاً ، تم تطبيق موديل هيدروديناميكي احادي البعد لنفاذ مياه البحر لغرض محاكاة التوزيعات الملحيه المعقدة بسبب التأثير المشترك للمصادر البرية والبحرية (الفصل 6) . النموذج يعتمد على سلسلة من البيانات لكل ساعه و لكامل العام 2014 . مع هذا الموديل تم تحليل تأثير السيناريوهات لإدارات مائيه مختلفة على التوزيعات الملحية وتحت ظروف مختلفة . حيث أظهرت النتائج الارتباط الوثيق بين تداخل مياه البحر وتصريف النهر . إن زيادة إستخدامات المياه عند المنبع وزيادة سحب المياه محلياً على طول النهر ستساهم في نفاذ مياه البحر وزيادة التراكيز الملحية على طول شط العرب . تحسين كمية ونوعية مصادر الأيرادات المائية يمكن أن يقلل من التراكيز الملحية ولكن بنسب مختلفه . تصريف المياه العائده من الأستخدامات البشرية المختلفه ، على الرغم من ملوحتها ، إلى داخل النهر تساعد في صد مياه البحر ، مع الأخذ بنظر الأعتبار إختيار أماكن تصريف المياه حيث إنها تؤثر على كل من مدى و توزيع الملوحة . كان التحليل الرقمي للسيناريوهات مفيداً بشكل خاص لدراسة التباين في التراكيز الملحية بالأتجاه الطولي في ظروف قاسية لأي من المتغيرات . و مع أسوأ السيناريوهات المفترضة ، فأن أفضل الستراتيجيات لإدارة المياه يمكن معاينتها والذي بدوره يتطلب تحليل المفاضلة بين استخدام المياه وملوحة المياه .

وقد تم تطوير موديلاً مركباً من نموذج الحل الأمثل لنظام متعدد الاهداف بالترابط مع نموذج المحاكاة المعد لهذا الغرض (الفصل 7) . حيث تمت محاكاة النظام الملحي بجميع مسبباته ، بما في ذلك مصادر الملوحة من المنبع، وتصريف المياه المستخدمه ونفاذ مياه البحر باستخدام الموديل الهيدروديناميكي الذي تم التحقق من صحته سابقاً و الذي يحاكي التوزيع الملحي على طول النهر لمختلف سيناريوهات التوزيعات المائيه . حيث تم فحص ستة منها بهذا الخصوص . أضهر الموديل القدره على تحديد أفضل الحلول التي تقلل كلاً من ملوحة مياه النهر والعجز في إمدادات المياه للاستخدام المنزلي واستخدامات الري . حيث تم استخدام الموديل للمفاضلة بين هذين الهدفين . إن الموديل المركب الذي يجمع بين نموذج المحاكاة والنموذج الأمثل يوفر المعلومات لصناع القرار بما يساعد في إدارة وتخفيف آثار الملوحة في المنطقة .

أضهرت النتائج المتحصله من إتباع المنهج المركب مع التبسيط المستخدم قلل نوعاً ما من تعقيد النظام المائي الى موديل احادي البعد والذي سهل الأداره والتحكم . كما ان قاعدة البيانات الجديده وخطوات التحليل المترتبه أنتجت أداة جديده لدعم إتخاذ القرارات التي ، مع مزيد من التحسينات ، يمكن أن تلائم سيناريوهات أكثر تعقيداً . وخلصت الدراسة (الفصل 8) بأن فهم التغيرات الملحيه ذات المستويات العاليه في نظم الدلتا المعقدة وعالية الديناميكية ، بأعتبار الاسس الصلده التي تم وضعها من خلال هذه الدراسة ، تلعب دوراً مركزيا في تصميم التدابير لضمان الأستخدام المستديم وإدارة الموارد المائيه الفعاله . شكلت قاعدة البيانات الشاملة والمفصلة الأساس للموديل التحليلي الذي يمكنه التنبؤ بأمتداد مياه البحر نسبتاً إلى مصادر الملوحة الأخرى عند مصب النهر ، وكذلك بناء الموديل الهيدروديناميكي الذي يمكنه التنبؤ بالتغيرات الملحية في النظام المائي تحت وطأة

الاستخدام المفرط . إن قدرة الموديلات المستخمه على التكيف للتعامل مع الظروف المتغيرة يجعلها قابلـة للتطبيـق بشكل مباشر مـن قبل المسؤوليـن عن إدارة الموارد المائيـه . ويمكـن تطبيق هذا الإجراء على أنظمـة مماثلة أخرى.

كمـا لاحضت الدراسه بأن الجهود الحالية في مجال إدارة الملوحة ليست كافية لمعالجة الأزمة المتصاعدة بالشكل الصحيح . وأن المراقبة المستمرة لنوعية المياه يمكن من خلالها تقييم الأثر نسبتاً لمصادر الملوحة المختلفة وفي أوقات مختلفة ، لا سيما مياه البحر ومصادر الاملاح المحلية . السيطره على امتداد اللسان الملحي والآثار المحلية يجب أن تأخذ في الأعتبار الاختلافات في نوعية وكمية مياه البزل وتصريف مياه الصرف الصحي على طول النهر ، وكذلك في كل من نهر دجلة و الفرات و الكرخة والكارون . ان الأزمه يمكن تفاديها فقط من خلال التعاون والتنسيق في إدارة المياه من قبل جميع الدول المتشاطئة ، والتي تتطلب تحولاً من النهج الحالي المبني على أساس التخطيط الأحادي الجانب لإدارة المياه إلى التعاون المشترك في إدارة الموارد المائية الدوليه . كما ان الدعم من قبل المجتمعات الإقليميه والدوليه يمكن أن يسهم في هذه النقلة النوعية . استراتيجيات التخفيف من الأزمة يجب أن تجد السبل لزيادة التدفقات من مصادر الأنهار وتحسين نوعية المياه . في نفس الوقت هناك حاجة إلى أتخاذ التدابير على المستوى المحلي لتجنب مياه الصرف المنزلية والصناعيـة والمياه عاليـة الملوحة من الأهوار الى مجرى النـهر . هذه الجهود ينبغي ان تكون مدعومه بالمعلومات العلمية الرصينه .

REFERENCES

Abbass, R.H, Abdul-Hussan J.K., and Resen A.K., 2014. Assessment of Water Quality of Shatt al-Arab River in north of Basra. Iraqi Journal of Aquaculture, 11(1): 37-56. (in Arabic)

Abdullah, D.A., Masih I., Van der Zaag P., Karim U.F.A, Popescu I., and Al Suhail Q., 2015. The Shatt al-Arab System under Escalating Pressure: a preliminary exploration of the issues and options for mitigation. International Journal of River Basin Management, 13 (2): 215–227 [doi:10.1080/15715124.2015.1007870].

Abdullah, A.D., Karim U.F.A., Masih I., Popescu I., van der Zaag P., 2016. Anthropogenic and tidal influences on salinity levels and variability of the Shatt al-Arab River, Basra, Iraq. International Journal of River Basin Management, 14(3):357-366 [doi:10.1080/15715124.2016.1193509].

Abdullah, A. D., Gisen, J. I. A., van der Zaag, P., Savenije, H. H. G., Karim, U. F. A., Masih, I., and Popescu, I., 2016. Predicting the salt water intrusion in the Shatt al-Arab estuary using an analytical approach, Hydrology and Earth System Science, 20:4031-4042 [doi:10.5194/hess-20-4031-2016].

Abdullah A., Popescu I., Dastgheib A., van der Zaag P., Masih I., Karim U., submitted. Analysis of possible actions to manage the longitudinal changes of water salinity in a tidal river. Submitted to Water Resources Management.

Abdullah A., Castro-Gama M.E., Popescu I., van der Zaag P., Karim U., Al Suhail Q., submitted. Optimization of water allocation in the Shatt al-Arab River when taking into account salinity variations under tidal influence. Submitted to Hydrological Sciences Journal.

Adib, A. and Javdan F., 2015. Interactive approach for determination of salinity concentration in tidal rivers (Case study: The Karun River in Iran). Ain Shams engineering Journal, 6: 785-793.

Afkhami, M., 2003. Environmental Effects of Salinity in the Karun-Dez basin, Iran. Seventh International Water Technology Conference Egypt 1-3 April, 229-233.

Afkhami, M., Shariat M., Jaafarzadeh N., Ghadiri H., and Nabizadeh R., 2007. Development a Water Quality Management Model for Karun and Dez Rivers. Iran. J. Environ. Health. Sci. Eng., 4(2): 99-106.

Aghdam, J. A. Zare M., Capaccioni B., Raeisi E., and Forti P., 2012. The Karun River waters in the Ambal ridge region (Zagros mountain Range, southwestern Iran): mixing calculation and hydrogeological implications. Carbonates Evaporites, 27: 251–267.

Ahmad, M.D., Islam Md. A., Masih I., Muthuwatta L. P., Karimi P., and Turral H., 2009. Mapping basin-level water productivity using remote sensing and secondary data in the Karkheh River Basin, Iran. Water International , 34(1):119-133.

Ajeel, S.G. and Abbas M.F., 2012. Diversity of Cladocera of the Shatt al Arab River Southern Iraq. Mesopot. J. Mar. Sci, 27(2):126-139.

Akanda, A., Freeman S., and Placht M., 2007. The Tigris-Euphrates river Basin: Mediating a Path towards Regional Water Stability. Al Nakhlah, Spring :63-74.

Al-Abaychi, J. and Alkhaddar R., 2010, Study on the Shatt al Arab River: The Latest Environmental Disaster. Report prepared to Iraq Institution for Economic Reform (IIER). (unpublished)

Al Amir, K.F., 2010. The Water Balancing in Iraq and Water Scarcity in the World. Baghdad: Dar Al Ghad. (In Arabic)

Al Aisa, S.A.A., 2005. Environmental study of aquatic plants and algae in the Shatt al-Arab River. PhD thesis, Agricultural collage, Basra University, Iraq. (in Arabic)

Alber, M., 2002. A Conceptual Model of Estuarine Freshwater Inflow Management. Estuaries, 25(6B):1246-1261.

Alcamo, J., Henrichs T., and Rosch T., 2000. World Water in 2025- Global Modeling and Scenario Analysis for the World Commission on Water for the 21st Century. Report A0002, Center for Environmental System Research, University of Kassel, Kurt Wolters Strasse 3, 34109 Kassal, Germany.

Alcamo, J., Doll P., Kaspar F., and Siebert S., 1997. Global Change and Global Scenarios of Water Use and Availability: An Application of WaterGAP1.0. Center for Environmental Systems Research (CESR), University of Kassel, Germany.

Al-Furaiji, M., Karim U., Augustijn D., Waisi B., and Hulscher S., 2015. Evaluation of water demand and supply in the south of Iraq, Journal of Water Resources and Desalination, DOI: 10.2166/wrd.2015.043.

Al Hajaj, M.M.K., 1997. Distribution of heavy elements in the water and sediments of Al-Ashar and Al-Kandak canals and determine their impacts on the algae. MSc thesis, Science collage, Basra University, Iraq. (in Arabic)

Allahyaripour, F., 2011. Study on the changes of the quality of the Karoun River at Khouzestan Plain in the View of Agricultural Usages and its Environmental Effects. Advances in Environmental Biology, 5(10): 3291-3295.

Al Mahmod, H.K.H., Al-Shawi A.J., Al-Emarah F.J.M., 2008. Assess the changes in some of the physical and chemical characteristics of the Shatt al-Arab River (1974-2005). Basra journal of agricultural science, 21:433-448. (in Arabic).

Al-Manssory, F.A., Abdul-Kareem M.A., and Yassen M.M., 2004. An Assessment of environmental pollution by Some Trace Metals in the Northern Part of Shatt al-Arab Sediments, Southern Iraq. Iraqi Journal of Earth Science, 4(2): 11-22.

Al-Meshleb, N.F., 2012. Systematic and ecological study of recent Ostracoda from Al Faw town, south of Iraq. Mesopot.J. Mar. Sci, 27(2):88-103.

Al Mudaffar-Fawzi, M.N. and Mahdi A. B., 2014. Iraq's Inland Water Quality and Their Impact on the North-Western Arabian Gulf. Marsh Bulletin, 9(1): 1-22.

Al-Saad, H.T., Farid, W.A. and Al-Adhub, A.Y., 2011. Uptake and depuration of water-soluble fractions (WSF) of crude oil by the bivalve Carbicula fluminea (Muller) from Shatt Al- Arab River. Mesopot. J. Mar. Sci, 26(2):134-145.

Al Sabah, B.J.J., 2007. Study the physic chemical behavior of mineral elements pollutions in the water and sediments of the Shatt al-Arab River. PhD thesis, Agricultural collage, Basra University, Iraq. (in Arabic).

Al-Tawash, B., Al-Lafta S. H., and Merkel B., 2013. Preliminary Assessment of the Shatt al Arab Riverine Environment, Basra Government, Southern Iraq. Journal of Natural Science Research, 3(13): 120-136.

Altinbilek , D., 2004. Development and Management of the Euphrates-Tigris Basin. Water Resources Development, 20(1):15-33.

Arabic Agricultural Statistics, 2002. Arab Agricultural Statistics, yearbook Vol.26. Arab Organization for Agricultural Development, Arab League. Available form: http://www.aoad.org/EAASYXX.htm [Accessed 5 October 2013].

Barros, M.T.L., Tsai F.T-C., Yang S-I., Lopes J. E. G., and Yeh W.W-G., 2003. Optimization of Large-Scale Hydropower System Operation, Journal of Water Resources Planning and Management, ASCE, 129(3):178-188.

Bates, C. B., Z.W. Kundzewicz, S. Wu and J.P. Palutikof, Eds., 2008: Climate Change and Water. Technical Paper of the Intergovernmental Panel on Climate Change, IPCC Secretariat, Geneva, 210 pp.

Beaumont, P., 1998. Restructuring of Water Usage in the Tigris-Euphrates Basin: The Impact of Modern Water Management Policies, Yale F&ES Bulletin, 103:168-186.

Becker, M.L., Luettich R.A., and Mallin M.A., 2010. Hydrodynamic Behaviour of the Cape Fear River and Estuarine: A Synthesis and Observational Investigation of Discharge - Salinity Intrusion Relationships. Estuarine, Coastal and Shelf Science, 88: 407-418.

Biedler, M., 2004. Hydropolitics of the Tigris-Euphrates River Basin with Implications for the European Union. CERIS Research Papers N1.

Brandimarte, L., Popescu I., Neamah N.K., 2015. Analysis of fresh-saline water interface at the Shatt Al-Arab estuary, International Journal of River Basin Management, 13 (1): 17-25.

Bobba, A. G., 2002. Numerical Modelling of Salt-water Intrusion due to Human Activities and Sea-level Change in the Godavari Delta, India. Hydrological Science, Special Issue: Towards integrated Water Resources Management for sustainable Development, 47: S67-S80.

Bucknall, J., 2007. Making the Most of Scarcity: Accountability for Better Water Management Results in the Middle East and North African MENA Development Report, the International Bank for Reconstruction and Development/ The World Bank, Washington, D.C.

Cai, H., Savenije H.H. G., and Gisen J. I. A., 2015.A coupled analytical model for salt intrusion and tides in convergent estuaries. Hydrological Sciences Journal, DOI:10.1080/02626667.2015.1027206.

Cai, H., Savenije H.H.G., Jiang C.,2014. Analytical approach for predicting fresh water discharge in an estuary based on tidal water level observation. Hydrology Earth System Science, 18:4153-4168.http://dx.doi.org/10.5194/hess-18-4153-2014.

Cai, H., Savenije H. H. G., Yang Q., Ou S., and Lei Y., 2012. Influence of River Discharge and Dredging on Tidal Wave Propagation: Modaomen Estuary Case. Journal of Hydraulic Engineering, 138: 885-896.

Cai, X., Ringler C., and Rosegrant M.W., 2006. Modeling Water Resources Management at the Basin Level, Methodology and Application to the Maipo River Basin. Research Report 149, International Food Policy Research Institute , Washington, D.C.

Canedo-Arguelles, M., Kefford J. B., Piscart C., Prat N., Schafer B. R., and Schulz C-J., 2013. Salinisation of rivers: An urgent ecological issue. Environmental Pollution, 173: 157-167.

Carrivick, J.L., Brown L.E., Hannah D.M., and Turner A.G.D., 2012. Numerical Modelling of Spatio-temporal Thermal Heterogeneity in a Complex River System. Journal of Hydrology, 414–415: 491–502.

Casulli, V. and Zanolli P., 2002. Semi-Implicit Modeling of Nonhydrostatic Free-Surface Flows for Environmental Problems. Mathematical and Computer Modelling, 36(9-10):1131-1149.

Chen, C. and Liu H., 2003. An unstructured Grid, Finite-Volume, Three-Dimensional, Primitive Equations Ocean Model: Application to Coastal and Estuaries. Journal of Atmospheric and Oceanic Technology, 20:159-186.

Chen X., Flannery M.S., and Moore D.L., 2000. Response Times of Salinity in Relation to Changes in Freshwater Inflows in the Lower Hillsborough River, Florida. Estuarine Research Federation, 23(5):735-742.

Das, A., Justic D., Inoue M., Hoda A., Huang H., and Park D., 2012. Impacts of Mississippi River Diversions on Salinity Gradients in a Deltaic Louisiana Estuary: Ecological and Management Implications. Estuarine, Coastal and Shelf Science, 111: 17-26.

Delft3D, 2012. Delft3D-FLOW, Simulation of Multi-Dimensional Hydrodynamics Flow and Transport Phenomena, Including sediments, User Manual Version: 3.15.20508, Deltares, delft, the Netherlands.

De Nijs, M.A.J. and Julie D. P., 2012. Saltwater Intrusion and ETM Dynamics in a Tidally-energetic Stratified Estuary. Ocean Modelling, 49–50:60–85.

De Voogt, K., Kite G., Droogers P., and Murray-Rust H., 2000. Modeling water allocation between wetlands and irrigated agriculture in the Gediz basin, Turkey. International of Water Resources Development, 16(4): 639-650.

Dickman, B. H., and Gilman M. J, 1989. Monte Carlo optimization. Journal of optimization theory and applications 60 (1): 149-157.

Dolphin, L., 2007.Water Population since Creation. World Population. online [access: 30-11-2012] http://donsnotes.com/reference/population-world.html.

El-Adawy, A., Negm A.M., Elzeir M.A., Saavedra O.C., El-Shinnawy I.A., and Nadaoka K., 2013. Modeling the Hydrodynamics and Salinity of El-Burullus Lake (Nile Delta, Northern Egypt). Journal of Clean Energy Technologies, 1 (2):157-163.

El-Kharraz, J., El-Sadek A., Ghaffour N., and Mino E., 2012. Water Scarcity and Drought in WANA Countries. Procedia Engineering, 33:14-29.

Falkenmark, L.M. Andersson R. Castensson and K. Sundblad 1999. Water, A Reflection of Land Use- options for Counteracting land and Water Mismanagement. NFR,Swedish Natural Science Research Council, Stockholm, Sweden: page 27.

FAO (Food and Agricultural Organization of the United Nations), 1985. Water quality for agriculture, Vol. 29 Rev. 1. FAO, Rome.

FAO (Food and Agriculture Organization of the United Nations), 2007. AQUASTAT database.

FAO (Food and Agriculture Organization of the United Nations), 2009. Irrigation in the Middle East in Figures, AQUASTAT survey-2008. Roma, Italy.

FAO (Food and Agricultural Organization of the United Nations), 2012a. The State of the World's Land and Water Resources for Food and Agriculture. online [15-11-2012] http://www.fao.org/fileadmin/templates/solaw/images_maps/map_3.pdf.

FAO (Food and Agricultural Organization of the United Nations), 2012b. The State of the World's Land and Water Resources for Food and Agriculture. online [23-11-2012] http://www.fao.org/nr/water/aquastat/climateinfotool/index.stm

FAO (Food and Agriculture Organization of the United Nations), 2013. FAO Corporate Document Repository. Available from: http://www.fao.org/docrep/006/Y4360E/y4360e06.htm [Accessed 12 November 2013].

Farthing, M.W., Fowler K.R., Fu X., Davis A., and Miller C.T., 2012.Effects of model resolution on optimal design of subsurface flow and transport problems. Advances in Water Resources, 38:27-37.

Fernandez-Delgado, C., Baldo, F., Vilas, C., Garcia-Gonzalez, D., Cuesta, J.A., Gonzalez-Ortegon, E., and Drake, P., 2007. Effects of the River Discharge Management on the Nursery Function of the Guadalquivir River Estuary (SW Spain). Hydrobiologia, 587:125-136.

Galbraith H., Amerasinghe P., and Huber-Lee A., 2005. The Effects of Agricultural Irrigation on Wetland Ecosystems in Developing Countries: A Literature Review. International Water Management Institute. Comprehensive Assessment Secretariat, CA Discussion Paper 1 Colombo, Sri Lanka.

GAP, 2012. Republic of Turkey, Ministry of Development, Southeastern Anatolia Project, Reginoal Development Administration. Latest situation of GAP [Online] http://www.gap.gov.tr/about-gap/latest-situation-of-gap [accessed 20 Nov. 2012].

Gisen, J.I.A. Savenija H.H.G., Nijzink R.C., and Wahab A.K. Abd., 2015a. Testing a 1-D analytical salt intrusion model and its predictive equations in Malaysian estuaries. Hydrological Sciences Journal, 60(1):156-172.

Gisen, J. I. A., Savenije, H. H. G., and Nijzink, R. C., 2015b. Revised predictive equations for salt intrusion modelling in estuaries, Hydrol. Earth Syst. Sci. Discuss., 12:739-770.

Geopolicity, 2010. Managing the Tigris-Euphrates Watershed: the Challenges facing Iraq. Dubai, United Arab Emirates (U.A.E).

Gong, Z-T., Sun B-Z., Xu Z-M., and Long A-H., 2005. The Rough Set Analysis Approach to Water Resources Allocation Decisions in the Inland River Basin of Arid Regions. Proceeding of the Fourth International Conference on Machine Learning and Cybernetics, Guangzhou, China.

Grego, S., Micangeli A., and Esposto S., 2004. Water purification in the Middle East crisis: a survey on WTP and CU in Basra (Iraq) area within a research and development program. Desalination, 165:73–79.

Guda, C., Scheeff E. D., Bourne P. E., Shindyalov I. N., 2001. A new algorithm for the alignment of multiple protein structures using Monte Carlo optimization. Pacific Symposium on Biocomputing. Vol. 6.

Hameed, H.A. and Aljorany Y.S., 2011. Investigation on Nutrient Behavior along Shatt Al-Arab River, Basra, Iraq. Journal of Applied Sciences Research, 7(8): 1340-1345.

Haro, D., Paredes J., Solera A., and Andreu J., 2012. A Model for Solving the Optimal Water Allocation problem in River Basin with Network Flow Programming When Introducing Non-Linearities. Water Resources Manage 26:4059-4071.

Hart, B.T., Bailey P., Edwards R., Hortle K., James K., McMahon A., Meredith C., and Swadling K., 1990. Effects of Salinity on River, Stream and Wetland Ecosystems in Victoria, Australia. Water Research 24(9):1103-1117.

Helland-Hansen, E., Holtedahl T. and Liye K. A., 1995. Enviromental Effects. Vol. 3. Hydropower Development. Norwegian Institute of Technology, Trondheim.

Hessari, B., Bruggeman A., Akhoond-Ali A., Oweis T., and Abbasi F., 2012. Supplemental irrigation potential and impact on downstream flow of Karkheh River Basin of Iran. Hydrol. Earth System Sciences Discussions, 9:13519–13536.

Horrevoets, A. C., Savenije H.H.G., Schuurman J. N., and Graas S., 2004. The Influence of River Discharge on Tidal Damping in Alluvial Estuaries, Journal of Hydrology, 294: 213–228.

Huckelbridge, K.H., Stacy M.T., Glenn E.P., and Dracup J.A., 2010. An integrated Model for Evaluating Hydrology, Hydrodynamics, Salinity and Vegetation Cover in a Coastal Desert Wetland. Ecological Engineering, 36:850-861.

Hussain, S.A., 2001. Organic pollution sources of inland water and the possibility to control and reuse. Mesopotamian journal of marine science, 16(1):489-505. (in Arabic)

ICARDA (International Center for Agricultural Research in the Dry Areas), 2012. Iraq Salinity Assessment, Managing Salinity in Iraq's Agriculture, Current state, Causes, and Impacts, An overview of the scope and scale of soil and water salinity in Central and Southern Iraq , Report 1: Situation Analysis, ICARDA, Australia.

IEA (International Energy Agency publication), 2012. World energy outlook, Part C, Iraq energy outlook, ISBN: 978 92 64 18084 0, France, Paris.

Iglesias A., Garrote L., Flores F., and Moneo M., 2007. Challenges to Manage the Risk of Water Scarcity and Climate Change in the Mediterranean. Water Resource Management, 21:775-788.

Ippen, A.T. and Harlemen D.R.F., 1961. One-Dimensional Analysis of Salinity Intrusion in Estuaries. Technical bulletin no.5. Committee on Tidal Hydraulics, Corps of Engineers, U.S. Army.

Isave, V.A. and Mikhailova M.V., 2009. The Hydrography, Evaluation, and Hydrological Regime of the Mouth Area of the Shatt al-Arab River. Water Resources, 36(4):402-417.

Italy-Iraq, 2006. New Eden Master plan for Integrated Water Resources Management in the Marshlands Area. Volume 1, Overview of Present Conditions and Current Use of the Water in the Marshlands Area, Iraq.

Jaafer, A.M., 2010. Quantitative and qualitative study of the plankton in water bodies southern Iraq. MSc thesis. Science collage, Basra University, Iraq. (in Arabic)

JAMAB., 1999. Comprehensive Assessment of National Water Resources: Karkheh River Basin. JAMAB Consulting Engineers in association with Ministry of Energy, Iran. (In Persian).

Jassby, A.D., Kimmerer W.J., Monismith S.G., Armor C., Cloern J.E., Powell T.M., Schubel J.R., and Vendlinski T.J., 1995. Isohaline Position as a Habitat Indicator for Estuarine Populations. Ecological Application, 5 (1):272-289.

Ji, Z.G., Morton M.R., and Hamrick J.M., 2001. Wetting and Drying Simulation of Estuarine Processes. Estuarine, Coastal and Shelf Science, 53(5):683-700.

Jones, C., Sultan M., Yan E., Milewski A., Hussein M., Al-Dousari A., Al-Kaisy S., and Becker R., 2008. Hydrologic Impacts of Engineering Projects on the Tigris-Euphrates System and Its Marshes. Journal of Hydrology, 353:59-75.

Karamouz, M., Szidarovszky F., and Zahraie B., 2003. Water Resources Systems Analysis, Lewis Publishers, CRC Press Company, United State.

King, J., Brown C., and Sabet H., 2003. A Scenario-Based Holistic Approach to Environmental Flow Assessments for Rivers. River Research and Application, 19: 619–639.

Kingsford R.T. and Thomas R. F., 1995. Arid Australia and Their Waterbirds: A 50-Year History of Decline. Environmental Management, 19(6): 867-878.

Kirchner J., Moolman J.H., du Plessis H.M., and Reynders A.G., 1997. Causes and management of salinity in the Breede River valley, South Africa. Hydrology Journal, 5(1):98–108.

Kliot N., 1994. Water Resources and Conflict in the Middle East. Routledge, London.

KHRP, 2002. Downstream Impacts of Turkish Dam Construction on Syria and Iraq: Joint Report of Fact-Finding Mission to Syria and Iraq jointly researched, written and published by Kurdish Human Rights Project. The Ilisu Dam Campaign, The Corner House.

Kolars, F. and Mitchell W.A., 1991. The Euphrates River and the Southeast Anatolia Development Project. Southern Illinois University, United States of America.

Kolars, J., 1994. Problem of International River Management: The Case of the Euphrates. Water Resources Management Series: 2, International Water of the Middle East from Euphrates-Tigris to Nile. Oxford University Press: Bombay Delhi Calcutta Madras.

Korzun, V.I., Sokolov A.A., Budyko M.I., Voskresensky K.P., Kalinin G.P., Konoplyantsev A.A., Korotkevich E.S., Kuzin P.S., and Lvovich M.I., 1978. World Water Balance and Water Resources of the Earth. UNESCO.

Koudstaal R., Rijsberman F. R., and Savenije H., 1992. Water and Sustainable Development. International Conference on Water and the Environment-Development Issues for the 21st Century, Dublin, Ireland 277-290.

Kourakos, G. and Mantoglou A., 2013. Development of a multi-objective optimization algorithm using surrogate models for coastal aquifer management. Journal of Hydrology, 479:13–23.

Labadie, J.W., 2004. Optimal Operation of Multireservoir Systems: State-of-the-Art Review, Journal of Water Resources Planning and Management, 93:93-111.

Lemly, A.D., Kingsford R.T., and Thompson J.R., 2000. Irrigated Agriculture and Wildlife Conservation: Conflict on a Global Scale. Environmental Management, 25(5):485-512.

Lesser, G.R., Roelvink J.A., van Kester J.A.T.M., and Stelling G.S., 2004. Development and validation of a three-dimensional morphological model. Coastal Engineering, 51: 883-915.

Liu, W., Hsu M., Wu C., Wang C., and Kuo A.Y., 2004. Modeling Salt Water Intrusion in Tanshui River Estuarine System-Case-Study Contrasting Now and then. Journal of Hydraulic Engineering, 130 (9): 849-859.

Liu, W.C., Hsu M-H., Wu C-R., Wang C-F., and Kuo A.Y.,, 2007. Modeling the Influence of River Discharge on Salt Intrusion and Residual Circulation in Danshuei River Estuary, Taiwan. ScienceDirect, 27:900-921.

Loucks, D.P., Stakhiv E. Z., and Martin L.R., 2000. Sustainable Water resources Management. Journal of Water Resources Planning and Management, 126:43-47.

Loucks, D.P. and van Beek E., 2005. Water Resources Systems Planning and Management. An Introduction to Methods, Models and Applications. United Nations Educationals, Scientific, and Cultural Organization, Paris.

MacKay, H.M. and Schumann E.H., 1990. Mixing and circulation in the sundays river estuary, South Africa. Estuarine, Coastal and Shelf Science, 31(2): 203-216.

MacQuarrie, P., 2004. Water Security in the Middle East, Growing Conflict over Development in the Euphrates-Tigris Basin. Thesis, M.Phil International Pease Studies, Trinity College, Dublin, Ireland.

Marin Science Center, 1991. Shatt al Arab elementary Scientific Studies, Al Basra University, Ministry of higher education and scientific research. (unpublished)

Margoni, S. and Psilovikos A., 2010. Sustainable Management of Agisama Lagoon-River Nestos delta-Using R.E.MO.S. Daily monitoring Data of Water Quality and Quantity parameters : Trends, assessments, and natural hazards for the years 2000–2002. Desalination, 250(1):287-296.

Marjanizadeh, S., Qureshi A.S., Turral H., and Talebzadeh P., 2009. From Mesopotamia to the third millennium: The historical trajectory of water development and use in the

Karkheh River Basin, Iran. IWMI Working Paper 135. Colombo, Sri Lanka: International Water Management Institute. doi:10.3910/2010.206.

Maser, M.D., Son M.O., and Yasser A.G., 2011. Assessing the Risks of Invasions of Aquatic Invertebrates in the Shatt Al-Arab River. Russian Journal of Biological Invasions, 2(2):120-125.

Masih, I., Ahmad M.D., Turral H., Uhlenbrook S., and Karimi P., 2009. Analysing stream flow variability and water allocation for sustainable management of water resources in the semi-arid Karkheh River Basin, Iran. Physics and Chemistry of the Earth, 34 (4-5): 329–340.

Masih, I., 2011. Understanding Hydrological Variability for Improved Water Management in the Semi-arid Karkheh Basin, Iran. PhD thesis. UNESCO-IHE Institute for Water Education and Delft University of Technology, Delft, the Netherlands.

Mayer, A. and Munoz-Hernandez A., 2009. Integrated Water Resources Optimization Models: An Assessment of a Multidisciplinary Tool for Sustainable Water Resources Management Strategies. Geography Compass, 3(3):1176-1195.

Mckinney, D. C. and Cai X. ,1996. Multiobjective Optimization Model for water Allocation in the Aral Sea Basin. (available online).

McKinney, D. C, Cai X., Rosegrant M. W., Ringler C., and Scott C. A., 1999. Modeling Water Resources Management at the Basin Level: Review and Future Directions. SWIM Paper 6, International Water Management Institute, Colombo, Sri Lanka.

Meijer, K. S., van der Krogt W.N.M., and van Beek E., 2012. A New Approach to Incorporating Environmental Flow Requirements in Water Allocation Modeling. Water Resources Manage, 26:1271-1286.

Merrett, S,Allan, J.A., and Lant, C., 2003. Virtual Water- the Water, Food, and Trade Nexus Useful Concept or Misleading Metaphor?, IWRA, water international, 28(1):4-11.

Mohammed, M.H., 2011. Modified method for the determination of cobalt (II) and copper (II) ions by adopting Schiff base complexes in water of Shatt al Arab River Mesopot.J. Mar. Sci, 26(2):170-181.

Moyel, M.S., 2014. Assessment of Water quality of the Shatt al Arab River, Using Multivariate Statistical Technique. Mesopotamia Environment Journal, 1(1): 39-46.

Myakisheva, N.V., 1996. The Influence of Seasonal and Year-to Year Variability of Water Discharge From the Lake Ladoga-Neva River System on the Salinity Regime of the Baltic Sea. Hydrobiologia 322:99-102.

NCWRM, 2009. National Centre for Water Resources Management, Ministry of Water Resource of Iraq, Annual Report, Baghdad. (unpublished)

Ng, S.M., Burling M., Rusbridge S. and Bailey M., 2013. Modelling bitterns discharges at dampier salt, port headland for application to coastal waters management [online]. In:

Australasian Port and Harbour Conference (14th : 2013 : Sydney, N.S.W.). Coasts and Ports 2013: 21st Australasian Coastal and Ocean Engineering Conference and the 14th Australasian Port and Harbour Conference. Barton, A.C.T.: Engineers Australia, 2013: 588-594.Availability: <http://search.informit.com.au/documentSummary;dn=827753670723491;res=IELENG > ISBN: 9781922107053.

Nguyen, A.D. and Savenije H.H.G., 2006. Salt Intrusion in Multi-Channel Estuaries: A Case Study in the Mekong Delta, Vietnam. Hydrology and Earth System Sciences, 10:743-754.

Nguyen, A.D., Savenije H.H.G., Pham D.N., and Tang D.T., 2008. Using Salt Intrusion Measurements to Determine the Freshwater Discharge Distribution Over the Branches of a Multi-Channel Estuary: The Mekong Delta Case. Estuarine, Coastal and Shelf Science, 77:433-445.

Nielsen, D.L., Brock M.A., Ress G.N., and Baldwin D.S., 2003.Effects of Increasing Salinity on Freshwater Ecosystems in Australia. Australian Journal of Botany, 51:655-665.

Paredes-Arquiola, J., Andreu-Alvarez J., Martin-Monerris M., and Abel S., 2010. Water Quantity and Water Quality Models Applied to the Jucar River Basin, Spain. Water Resources Manage, 24:2759-2779.

Parsons, A.J. and Abrahams, A. D. 1994. Geomorphology of desert environments. Springer Netherlands.

Pereira, L.S., Cordery I., and Iacovides I., 2002. Coping With Water Scarcity. International Hydrological Programme, Technical Documents in Hydrology No.58, UNESCO, Paris.

Peters, N.E. and Meybeck M., 2000. Water Quality Degradation Effects on Freshwater Availability: Impact of Human Activities. International Water Resources Association, Water International, 25(2):185-193.

Prandle, D., 1985. On Salinity Regimes and the Vertical Structure of Residual Flows in Narrow Estuaries. Estuarine, Coastal and Shelf Science, 20: 615-635.

Quinn, N.W.T., 2011. Adaptive Implementation of Information Technology for Real-time, Basin-Scale Salinity Management in the San Joaquin Basin, USA and Hunter River Basin, Australia. Agricultural Water Management, 98:930–940.

Rahi, K.A. and Halihan T., 2010. Changes in the Salinity of the Euphrates River system in Iraq. Reg Environ Change, 10: 27-35.

Reinert T.R. and Peterson J.T., 2008. Modeling the Effects of Potential Salinity Shifts on the Recovery of Striped Bass in the Savannah River Estuary, Georgia–South Carolina, United States. Environmental Management, 41(5):753-765.

Risley, J.C., Guertin D.P., and Fogel M.M., 1993. Salinity Intrusion Forecasting System for Gambia River Estuary. Journal of Water Resources Planning and Management, 119: 339-352.

Roos J.C. and Pieterse A.J.H., 1995. Salinity and Dissolved Substances in the Vaal River at Balkfontein, South Africa. Hydrobiologia, 306:41-51.

Sakalauskas, L., 2000. Nonlinear Stochastic Optimization by the Monte-Carlo Method. Informatic, 11(4):455-468.

Salarijazi, M., Akhond-Ali A-M., Adib A. and Daneshkhah A., 2012. Trend and change-point detection for the annual stream-flow series of the Karun River at the Ahvaz hydrometric station. African Journal of Agricultural Research, 7(32): 4540-4552.

Savenije, H.H.G., 1986. One-Dimensional Model for Salinity Intrusion in Alluvial Estuaries. Journal of Hydrology, 85: 87-109.

Savenije, H.H.G., 1989.Salinity Intrusion Model for High-Water Slack, Low Water-Slack, and Mean Tide on Spread Sheet. Journal of Hydrology, 107: 9-18.

Savenije H.H.G., 1993. Predictive Model for Salt Intrusion in Estuaries. Journal of Hydrology, 148(1-4):203-218.

Savenije, H.H.G., 2005. Salinity and tides in alluvial estuaries. Amsterdam: Elsevier.

Savenije, H.H.G., 2012. Salinity and tides in alluvial estuaries [online]. Completely revised 2nd edition, Delft University of Technology, Available from: http://salinityandtides.com [Accessed 20 Jan 2015].

Savenije, H., Cai H., and Gisen J., 2013. Developing a coupled analytical model for analyzing salt intrusion in alluvial estuaries. AGU Fall Meeting Abstracts, 1: P 1664.

Seckler, D., Amarasingle U., Molden D., de Silva R., and Barker R. , 1998. World Water Demand and Supply, 1990 to 2025: Scenarios and Issues. International Water Management Institute (IWMI), Research Report 19, Colombo, Sri Lanka.

Simeonov, V., Stratis J.A., Samara C., Zachariadis G., Voutsa D., Anthemidis A., Sofoniou M., and Kouimtzis TH., 2003. Assessment of the Surface Water Quality in the Northern Greece. Water Research, 37: 4119-4124.

Shamout M. N. and Lahn G., 2015. The Euphrates in Crisis: Channels of Cooperation for a Threatened River. The Royal Institute of International Affairs, Chatham House. London.

Shao, W., Yang D., Hu H., and Sanbongi K., 2009 . Water Resources Allocation Considering the Water Use Flexible Limit to Water Shortage-A Case Study in the Yellow River Basin of China. Water Resources Manage, 23:869-880.

Shayal, A.H., 2010. The Role of the Arabs in the Construction of Irrigation Channels. Surmnraa Journal , 22(6): 253-270. (in Arabic)

Shapland, G., 1997. River of Discord International Water Dispute in the Middle East. HURST &COMPANY, London.

Shiati, K., 1991, A Regional approach to Salinity Management in River Basins, A Case Study in South Iran. Agricultural Water Management, 19:27-41.

Shiklomanov, I.A., 1998. World Water Resources: An Appraisal for the 21st Century. International hydrologic Program Report, UNESCO, Paris.

Sklar, F. H. and Browder J. A., 1998. Coastal Environment Impacts Brought about by Alteration to Freshwater Flow in the Gulf of Mexico. Environmental Management, 22(4): 547-562.

Singh, A.,2012. An overview of the optimization modelling applications. Journal of Hydrology, 466-467:167-182.

The Iraq Foundation, 2003. Draft Report Physical Characteristics of Mesopotamian Marshlands of Southern Iraq, Background Material Prepared for the Technical Advisory Panel, Eden Again Project.

The World Bank, 2014. Population Growth Data. Available from: http://data.worldbank.org/indicator/ SP.POP.GROW [Accessed 2 May 2014].

Thomas, G.A. and Jakeman, A.J., 1985. Management of Salinity in the River Murray Basin. Land Use Policy, 2(2):87-101.

UNDP (United Nations Development Programs), 2006. Water Scarcity, Risk and Vulnerability, Human Development Report, New York, USA.

UNEP (United Nations Environmental Program), 2001. The Mesopotamian Marshlands: Demise of an Ecosystem, Early Warning and Assessment Technical Report, UNEP/DEWA/TR.01-3 Rev.1. Division of Early Warning an Assessment. Nairobi, Kenya.

UN-ESCWA and BGR (United Nations Economic and Social Commission for Western Asia; Bundesanstalt für Geowissenschaften und Rohstoffe), 2013. Inventory of Shared Water Resources in Western Asia, Beirut.

United Nations, 2011. Population Distribution, Urbanization, Internal Migration and Development: An International Perspective. Department of Economic and Social Affairs, Population Division.

UNWP (United Nations White Paper), 2011. Managing Change in the Marshlands: Iraq's Critical Challenge. Report of the United Nations Integrated Water Task Force for Iraq, United Nations.

UN Water, 2012. Statistics. Water Resources. online [access: 30-11-2012] http://www.unwater.org/index.html.

UN Water and FAO (Food and Agricultural Organization of the United Nations), 2007. Coping With Water Scarcity, Challenge of the Twenty-First Century. World Water Day, 22nd March.

Van Breemen, M.T.J., 2008. Salt intrusion in the Selangor Estuary in Malaysia. MSc thesis. University of Twente, Enschede, the Netherlands.

Van den Heuvel, S., 2010. Modeling the hydrodynamics and salinity of the Pontchartrain Basin. MSc thesis. Delft University of Technology, Delft, the Netherlands.

Van der Burgh, P., 1972. Ontwikkeling van een methode voor voorspellen van zoutverdelingen in estuaria, kanalen en zeeen. Rijkswaterstaat Rapp. 10-72.

Van der Zaag, P., 2007. Asymmetry and Equity in Water Resources Management; Critical Institutional Issues for Southern Africa. Water Resources Management , 21:993-2004.

Varis, O., Kummu M., and Salmivaara A., 2012. Ten Major River in Monsson Asia-Pacific: An Assessment of Vulnerability. Applied Geography, 32:441-454

Vaz, N., Dias J.M., Leitao P., and martins I., 2005. Horizontal Patterns of Water Temperature and Salinity in an Estuarine Tidal Channel: Ria de Aveiro. Ocean Dynamics, 55: 416–429.

Vaz, N., Dias J.M., and Leitao P.C., 2009. Three-Dimensional Modelling of a Tidal Channel: The Espinheiro Channel (Portugal). Continental Shelf Research, 29: 29–41.

Vörösmarty, C. J., Green P., Salisbury J., and Lammers R.B., 2000. Global Water Resources: Vulnerability From Climate Change and population Growth. Science, 289:284-288.

Vörösmarty,C.J. and Sahagian, D., 2000. Anthropogenic Disturbance of the Terrestrial Water Cycle. BioScience, 50 (9): 753-765.

Wang, Y. and Peng H., 2009. A integrated Water Quantity-Quality Method for Water Resources Management. International Conference on Environmental Science and Information Application Technology, 178-181.

Wang, Y., Liu Z., Gao H., Ju L., and Guo X., 2011. Response of Salinity Distribution around the Yellow River mouth to Abrupt Changes in River Discharge. Continental Shelf Research, 31: 685-694.

Weifeng, Y. And Chesheng, Z., 2010. Study on Sustainable Utilization Coupled Management Model for Water Resources in an Arid irrigation District (China). The Natural Science Foundation of China and Open Fund of State Key of Water resources and hydropower Engineering Science (Wuhan University).

Whitney, M.M., 2010. A Study on River Discharge and Salinity Variability in the Middle Atlantic Bight and Long Island Sound. Continental Shelf Research, 30: 305-318.

WHO (World Health Organization), 1996. Guidelines for drinking-water quality, Health criteria and other supporting information, 2nd ed. Vol. 2. World Health Organization, Geneva.

Windsor, J. S. ,1973. Optimization Model for the Operation Flood Control Systems, Water Resources Research, 9(5):1219-1226.

Wong, K.C. , 1995. On the relationship between long-term salinity variations and river discharge in the middle reach of the Delaware estuary, J. Geophys. Res., 100(C10), 20705–20713, doi:10.1029/95JC01406.

World Bank, 2006. Making the Most of Scarcity, Accountability for Better Water Management Result in the Middle East and North Africa: MENA Development report, The World Bank, Washington, D.C.

Xinfeng, Z. and Jiaquan D., 2010. Affecting Factors of Salinity Intrusion in Pearl River Estuary and Sustainable Utilization of Water Resources in Pearl River Delta.Alliance for Global Sustainablility Bookseries, 18:11-17.

Xue, P., Chen C., Ding P., Beardsley R. C., Lin H., Ge J., and Kong Y., 2009. Saltwater Intrusion into the Changjiang River: A Model-Guided mechanism Study. Journal of Geophysical Research, 114:1-15.

Xu, M., Van Overloop P.J., and Van de Giesen N.C., 2011. Model Selection for Salt Water Intrusion in Delta Areas. 25th ICID European Regional Conference, Integrated Water Management for Multiple Land Use in Flat Coastal Areas, Groningen, The Netherlands, 16-20 May. paper II-17.

Yang, Z. F., Sun T., Cui B.S., Chen B., and Chen G.Q., 2009. Environmental Flow Requirements for Integrated Water Resources Allocation in the Yellow River Basin, China. Communication in Nonlinear Science and Numerical Simulation, 14:2469-2481.

Yeh, W.W-G., 1985. Reservoir Management and Operation Models: A State-of-the-Art Review, Water Resources Research, 21:1797-1818.

Zhang, E., Savenije H.H.G., Wu H., Y. Kong, and J. Zhu, 2011. Analytical solution for salt intrusion in the Yangtze Estuary, China. Estuarine, Coastal and Shelf Science, 91:492-501.

Zhang, E.F. , Savenije H.H.G., Chen S.L., and Mao X.H., 2012. An analytical solution for tidal propagation in the Yangtze Estuary, China. Hydrology and Earth System Sciences, 16: 3327-3339.

Zhichang, M, Huanting S., Liu J.T., and Eisma D., 2001. Types of Saltwater Intrusion of the Changjiang Estuary. Science in China (series B) 44(1 Supp):150-157.

ACKNOWLEDGMENTS

Firstly, I thank my Almighty God, my good Father, for letting me through all the difficulties and granting me the wisdom, health and strength to undertake this research task. I have experienced Your guidance and provision every resource required for success. You are the one who enable me complete this study and let me finish my degree. I will keep on trusting You for my future. Thank you, Lord.

I would like to express my special appreciation and thanks to my research guide Professor Pieter van der Zaag for the continuous support of my PhD study and related research, for his patience, motivation, and immense knowledge, and for the valuable guidance, scholarly inputs and consistent encouragement I received throughout the research work. This feat was possible only because of the unconditional support provided by you. You have been a tremendous mentor for me. I would like to thank you for encouraging my research and for allowing me to grow as a research scientist. Your advice on both research as well as on my career have been priceless. Your guidance helped me in all the time of research and writing of this thesis. I could not have imagined having a better advisor and promoter for my PhD study.

Besides my advisor, I would like to thank my co-promoter and the rest of my supervisors: Professor Usama Karim, Dr. Ioana Popescu, Dr. Ilyas Masih, Professor Qusay Al Suhail, and Dr.Nadia Fawzi, for their insightful comments and encouragement, but also for the hard question which incepted me to widen my research from various perspectives. Persons with an amicable and positive disposition, they have always made themselves available to clarify my doubts despite their busy schedules and I consider it as a great opportunity to learn from their research expertise. I appreciate all their contributions of time, ideas, experience productive and stimulating. In particular, I am grateful to Prof. Usama Karim for enlightening me the first glance of research, and for the excellent preparation at the first stage of this project, with it was possible to provide and install the monitoring equipments in Iraq. Thank you all, for all your help and support.

For this dissertation I would like to express my sincere gratitude to my committee members, Professor Nadhir Al-Ansari, Professor Flavio Martins, Professor Hubert Savenije, and Professor Eelco van Beek for serving as my committee members and for their time, interest, and helpful comments. I also want to thank you for letting my defense be an enjoyable moment, and for your brilliant comments and suggestions, thanks to you.

For the analytical approach, I would especially like to thank Professor Hubert Savenije and Dr. Jacqueline Isabella Gisen on their support testing and applying the 1-D analytical salt intrusion model. In my work of developing the hydrodynamic model, I am particularly indebted to Dr. Ali Dastgheib for all his support, comments, and discussions, for his valuable suggestions and concise comments on some of the research papers of the thesis. I also thank Dr. Mick van der Wegen, and Johan Reyns for inspirational discussions regarding the hydrodynamic model. For the optimization work, I am especially grateful for conservations with Mario Castro Gama as we strived to write the final code. I thank him for his support and contribution. My sincere thanks also go to Dr. Ali Dastgheib for his support calling the Delft3D model by the MATLAB software.

I am thankful for the ever available support and kind interaction with many UNESCO-IHE colleagues. I remember the staff of the PhD fellowship department, particularly Jolanda Boots, Computer Centre, Library (the center of learning resources), and social and culture department I acknowledge and appreciate them for all their helps and services. Many thanks to Susan Graas for her valuable help in translating summary into Dutch. Special thanks to Jos Bult, Selda Akbal, and Gerda de Gijsel for their cooperation and support. Some members of the Institute have been very kind enough to extend their help at various phases of this research, whenever I approached them, and I do hereby acknowledge all of them. The interaction with many people from various cultures and nationalities at UNESCO-IHE was a unique and highly enriching experience, which has brought added respect for the diversity and difference of opinion and cultures in my life.

I gratefully acknowledge the funding sources that made my Ph.D. work possible. Completion of this doctoral dissertation was possible with the financial support of the

Higher Communities for Developing Education in Iraq (HCED). I would also like to acknowledge the Ministry of Water Resources of Iraq and Marine Science Center at Basra, Iraq for their support conducting the field measurement and providing required information. I have appreciated the collaboration of other competent authorities including environmental, agricultural, and water distribution departments. Special thanks to Mohammed Rheef for providing the cover picture of this dissertation. My field study was made less obstacle ridden because of the presence of their supports and cooperation.

Last but not the least; I would like to thank my family, my brothers and sisters for supporting me spiritually throughout doing this research and my life in general. A special thanks to my mother for all her love and encouragement. Words cannot express how grateful I am to my mother and my sons for all of the sacrifices that you've made on my behalf. Your prayer for me was what sustained me thus far. I would also like to thank all of my friends who supported me in writing, and incented me to strive towards my goal. Thank you.

ABOUT THE AUTHOR

Ali Dinar Abdullah was born in 1977 in Missan, Iraq. In 1998 he obtained his BSc degree in Civil Engineering from Basra University, in Basra. He completed his MSc degree in Water Resources Management in 2007 at UNESCO-IHE Institute for Water Education, Delft, the Netherlands. His master research was on the design of operating rules for complex reservoir systems, applying artificial neural networks.

Ali has over ten years of experience working in the field of land and water development. He joined the Ministry of Water Resources of Iraq in 2001. Within this ministry he worked mainly on irrigation and drainage projects. He was involved in design, implementation, and rehabilitation of several projects such as land reclamation, irrigation and drainage schemes. He also participated in an NGO-supported programme to improve water and environmental sanitation services, particularly in rural communities. He has extensive experience on issues related to providing drinking water and sanitary facilities and engaging with communities to help them to develop their own solutions to local problems.

In 2008 he became head of maintenance of water resource projects department at the Ministry of Water Resources, and mostly worked on the maintenance of flood protection works, water control and distribution structures, irrigation and drainage networks and main waterways.

In September 2012, Ali returned to Delft to embark on his PhD studies at UNESCO-IHE Institute of Water Education. His PhD research focuses on issues of water availability and salinity changes for improved water resources management in the Shatt al-Arab River, Iraq. During his research he was responsible to organize and conduct a comprehensive data collection campaign in a hitherto unsurveyed river, under difficult climatic and security conditions. Ali is the first author of a number of scientifically important and practically relevant papers and has presented his research at national and international workshops.

Ali is married to Shaymaa and they have two daughters Fatimah and Malk, and a son Hasan.

List of publications

Abdullah, D.A., Masih I., Van der Zaag P., Karim U.F.A, Popescu I., and Al Suhail Q., 2015. The Shatt al-Arab System under Escalating Pressure: a preliminary exploration of the issues and options for mitigation. International Journal of River Basin Management, 13 (2): 215–227 [doi:10.1080/15715124.2015.1007870]

Abdullah, A.D., Karim U.F.A., Masih I., Popescu I., van der Zaag P., 2016. Anthropogenic and tidal influences on salinity levels and variability of the Shatt al-Arab River, Basra, Iraq. International Journal of River Basin Management 14(3):357-366, [doi:10.1080/15715124.2016.1193509]

Abdullah, A. D., Gisen, J. I. A., van der Zaag, P., Savenije, H. H. G., Karim, U. F. A., Masih, I., and Popescu, I., 2016. Predicting the salt water intrusion in the Shatt al-Arab estuary using an analytical approach, Hydrology and Earth System Science, 20:4031-4042 [doi:10.5194/hess-20-4031-2016].

Abdullah A., Popescu I., Dastgheib A., van der Zaag P., Masih I., Karim U., submitted. Analysis of possible actions to manage the longitudinal changes of water salinity in a tidal river. Submitted to Water Resources Management.

Abdullah A., Castro-Gama M.E., Popescu I., van der Zaag P., Karim U., Al Suhail Q., submitted. Optimization of water allocation in the Shatt al-Arab River when taking into account salinity variations under tidal influence. Submitted to Hydrological Sciences Journal.